to Monica

Come walk on the

Wild Side with me and

enjoy the beauty of these

precious orchids

Best regards

Connie Braun

1 3/06

Wild Love Affair

Essence of Florida's Native Orchids

Photography and text by
Connie Bransilver

Foreword by Eric Hansen
Preface by Stuart Pimm, Ph.D.
Essays by Paul Martin Brown and John Beckner

WESTCLIFFE PUBLISHERS

westcliffepublishers.com

Individual butterfly orchid (Encyclia tampensis) *plants are highly variable in size and color. The more typical coloration displays yellowish or copper sepals and petals with a white-veined lip and pink spot in the middle.*

First Frontispiece: *Cigar or bee-swarm orchid* (Cyrtopodium punctatum)
"Orchids are the special treats—eye candy of the swamp. They still exist in profusion in isolated pockets despite poaching and the effects of water deprivation."
                                                    —*Connie Bransilver and Larry W. Richardson*
                                                    Florida's Unsung Wilderness: The Swamps

Second Frontispiece: *Fakahatchee ladies'-tresses* (Sacoila lanceolata *var.* paludicola) *appears to be native only to Collier County, Florida. This scarlet, beaked terrestrial orchid is a striking jewel among the verdant foliage of Fakahatchee's trams and hammocks.*

To Mother, a shade lily, and to Pop, who made me do it.

"God is not hiding from us. He is speaking to us through flowers, through our thoughts, through all creation."

—*Paramahansa Yogananda*

The bold, open sensuality of large, wild orchids like night-scented orchids (Epidendrum nocturnum) is an in-your-face dynamic force—alive, confusing, and evocative of impropriety.

Author's Note

"We live in a world of appearances and illusions, where little is certain. . . . To a substantial extent, this has always been the case—and the history of the human mind . . . is nothing if not a history of coming to terms with this ambiguity. [The Eastern tradition] simply accepts that we are wrapped in a veil of illusion . . . and that the goal of the human spirit is to one day pass through that veil into clarity."

—Animal *by James Balog*

Wild orchids, mere plants, open a door into wilderness, and for me, it is a door into a parallel universe, a time of no-time where my spirit can soar. As I learn more and more, as orchids unveil their secrets and those of the wild places where they live, I am transformed. I am not only at one with the land and water and sky, I am also at peace with myself, strong and serene.

These elusive beauties and the mystique surrounding them reconnect me with my youth and the freedom I felt nearly every day as I rode my horses on the mesas of New Mexico or through the dark and damp cottonwood bosques. I dared then to be beautiful, to drink the fresh air and ride hard. In the jungle I can dare again, be less obedient, and explore wilderness where I find it.

Exploration of the outer world takes on a deeper meaning if it also pushes us to explore our inner world. In my photographs and in my writing, I try to move beyond the confines of perfect focus and into the realm of the spirit, for in life things are never completely clear. Realists may not like my photos. Dreamers will. They are all deeply personal and represent huge pieces of my own soul. Life—its exuberance—is central to my being, and I want to share my awe with those who care to see it.

I hope that if I tell a love story between humans and wild orchids, I might draw people across the boundary, away from themselves and into the realization that every species has the right to exist, that we are all in this together—diminishing one diminishes us all. Growth in understanding always comes with a certain level of discomfort, for exposure to and grappling with different ways of viewing existence is sometimes precarious as life poses new questions and challenges. But greater understanding should not diminish our awe, it should expand and enhance it.

Join me as I continue to search for my own spiritual connection to wildness, for the link between humans and their environment, and for the emotional involvement that precedes love, respect, and commitment to a landscape and to a place. Experience the joy of feeling deeply—the joy of passion. Dare to have a wild love affair.

—Connie Bransilver
Artemis Images, www.ConnieBransilver.com

Billy Ghost orchid (Dendrophylax lindenii) *nods on his long stalk, trailing his extended four-to-six-inch nectar spur. The unusual second flower is still in bud. Rare now, ghost orchids are found only in the swamps of southwestern Florida, and, at one time at least, were found in similar habitats in Cuba.*

Contents

Foreword

The first time I saw a wild orchid in bloom was deep in the Borneo rainforest in 1992. It was *Paphiopedilum sanderianum*, the legendary species of almost indescribable beauty and romantic appeal to rare plant collectors, orchid breeders, botanists, taxonomists, and conservationists alike. I found the plant on a remote limestone cliff after traveling upriver by dugout canoe and on foot for six days.

In the years that followed, I counted and studied the orchids on these limestone hills in an attempt to understand why they thrived in this small corner of the world. I also wanted to help save the plants from illegal collection by creating a plant salvage operation. It would allow local Berawan and Punan villagers to cooperate with a Malaysia-based orchid nursery to gather breeding stock, propagate seedlings, and then sell these highly valued plants to collectors worldwide. This sustainable, grass-roots, self-funding conservation project would have provided meaningful work to local villagers with few employment opportunities while helping to save a threatened plant species. What I didn't realize at the time was that international conservation laws prevented such a project.

According to international treaty, the collection of this species is prohibited because of its rare and endangered status. It is perfectly legal to log, mine, quarry, burn, or flood the habitat of rare and endangered orchid species for development, but the collection of these same plants for scientific study or commercial propagation is punishable by huge fines and prison sentences—all in the name of conservation.

Within ten years, this rare plant and thousands more like it were destroyed. The limestone

cliffs where the orchids grew in profusion were clear-cut—legally. The naked rock hills were then blasted, quarried, and pulverized into gravel for a small airport runway close by, just across the Melinau River from Mulu National Park, a World Heritage Site. Visitors come from all over the world to visit the park, and, more specifically, to view this rare orchid. I doubt most of them realize how many thousands of *Paphiopedilum sanderianum* were lost on the other side of the river in the effort to provide easy access to the park.

My experiences in Borneo are not unique. Habitat is being threatened worldwide, with no provisions to protect or move plant species. In response to all of this, I decided to write my book *Orchid Fever*, which in part challenges the ways laws are being interpreted and implemented by a select few people who have managed to work around them for the sole purpose of obtaining exclusive use of rare plant material for research and commercial use. I also began accepting invitations to give lectures at botanical institutions across the United States and at the World Orchid Conference in Malaysia, about orchid conservation, plant salvage, and sustainable *ex situ* orchid propagation.

After presenting a talk at Marie Selby Botanical Gardens in Sarasota, Florida, I fell into conversation with Connie Bransilver, who had been sitting in the audience. We discussed the plight of rare and endangered orchid species of Florida, and she described how habitat destruction in the form of urban sprawl, agriculture, road building, and water diversion was threatening the Everglades and Fakahatchee Strand—a place then unknown to me. I thanked her for attending the lecture, we exchanged cards, and she disappeared into the crowd. A year later, I received an invitation to speak at the Conservancy of Southwest Florida. The requested topic was orchid conservation and, of course, the invitation was from Connie. I gave my talk and two days later found myself waist deep in the Fakahatchee Strand surrounded by alligators, while the diminutive Connie, who was nearly submerged, calmly took photos of a flowering ghost orchid growing on a tree.

A dozen baby alligators basked in a patch of sunlight on a partially exposed log. They made a cute sort of squeaking noise, and out of curiosity I moved through the water to get a closer look at them. When I was about 20 feet from the log, the babies turned their heads in my direction and stopped squeaking. They paused, examined me briefly, and then started to make a different sort of sound.

"That's probably close enough." Connie said over her shoulder. "Back off a little bit. But slowly." She went back to her photography without saying another word.

We waded deeper into the cypress swamp, and as the sunlight dimmed, the rest of the world dropped away. I began to realize how at ease Connie was in this habitat. It was as if she had been moving through these trees and sloughs all of her life. I, too, felt at home in Fakahatchee Strand. I was reminded of the natural rhythms and sense of vulnerability, timelessness, and serenity that I have experienced in Borneo's rain forests, where humans are, at best, temporary visitors, reassigned to their proper place in the order of things. We couldn't have been more than a half mile from the road, but we were definitely in another world—a special place—unique, but somehow familiar.

For several hours Connie showed me clamshell orchids, pond apples, ghost orchids, duck weed,

cigar orchids, vanilla, butterfly orchids, the tiny *Epidendrum anceps*, medicinal plants used by the Calusa Indians, bromeliads, and countless other jungle plants. She told me their stories, bringing this strange and wonderful place to life as a cohesive and intricate whole. I learned about aquifers, rainfall patterns, sloughs, strands, and the importance of unfettered water flow to the complex habitats it sustains.

It wasn't until we were out of the swamp and drying our clothes that Connie told me the baby alligators I'd been investigating were on the verge of calling to their mother for help.

"That's something you don't want to have happen out there," she cautioned. "Mother alligators don't ask questions first, and it's difficult to run in four feet of water."

The Fakahatchee Strand is now a real place for me, populated by showy orchids and other epiphytes, as well as alligators, swamp turkeys, great egrets, green herons, little blue herons, roseate spoonbills, and mosquitoes. The sounds and smells and shifting patterns of light are still with me, as are the threats to this beleaguered and diminishing patch of natural wonder. Now, 3,000 miles away from Florida's swamps, I am filled with gratitude for people like Connie, who understand our earth's environmental heritage and fragile ecosystems and who are willing to step forward to protect and draw attention to the bounty of these few and precious places.

—Eric Hansen

Eric Hansen is the author of Orchid Fever: A Horticultural Tale of Love, Lust, and Lunacy

Preface

Each winter, visitors from Europe, the northern U.S., and beyond arrive in Florida to snorkel off our coral reefs. Their white skin the color of cooked lobsters, they clamber back into the boat and exclaim in a dozen different languages about the fish and corals they have seen. Others venture into the Everglades to watch wildlife; still others explore the myriad natural areas of our state. More adventurous cousins visit the Amazon, or Costa Rica, or even wilder places. Pallid relatives back home only have their carefully tended houseplants and tanks of tropical fish to bring color to their lives.

One such tourist traveling in Brazil wrote to his father, peppering his letters with "striking," "luxuriant," "no person could imagine anything so beautiful," and "exquisite glorious pleasure." He penned that the diversity of life "bewilders the mind. If the eye attempts to follow the flight of a gaudy butterfly, it is arrested by some strange tree or fruit; if watching an insect, one forgets it in the stranger flower it is crawling over . . . ." In the forest, he enthused about "the elegance of the grasses, the novelty of the parasitical plants, the beauty of the flowers, the glossy green of the foliage . . . . the noise of the insects." Charles Darwin's letters go on and on like this for pages. Especially when one is escaping an English winter, the warm parts of the New World are a sensory shock.

Of course, young Darwin did more than write letters home. On his long voyage, he thought at length about what he had seen and experienced. On the way home, his ship, HMS *Beagle*, stopped off at the Galapagos Islands. Struck by the fact

that each island had distinct (if broadly similar) kinds of tortoises, mockingbirds, and small black finches, he dubbed the creatures "aboriginal productions." He realized that these islands were places where "species are born."

Darwin did not visit Florida and the Caribbean. Had he done so, he could have developed his world-changing ideas here as easily he did in the Galapagos. However, there are only a few places in the world that are special enough to have generated his ideas. These are the places that are not only rich in species, but in endemics— species that are found nowhere else. It is not only the richness of species of fish and butterflies, birds and flowers, that bedazzles our state's visitors (and its residents too) but their uniqueness.

There is a dark side to Darwin's story. The very uniqueness of endemic species makes them vulnerable to human impacts. As we clear forests, drain wetlands, and otherwise modify natural areas, we are poised to place a third or more of the variety life on an inexorable path to extinction within a generation or so. Whether one calls that variety biodiversity, the product of evolution, or God's Creation, such losses are unacceptable.

So what can we do?

To save the variety of life, we must both catalogue it and map it. We must know what biodiversity we have and where it lives. So armed, we can set the best priorities for conserving the special places where the most vulnerable species live. The problem is that we do not have names for most species (perhaps 90% of all species lack names) and we know the geographical distributions of only a small fraction of the species for which we do have names. By the time we complete the catalogue of life on earth, it will be too late to save many of these species.

We must act now, on what we know now. We *do* know orchid names. For scientist and nonscientist alike, the exuberance of orchid colors and shapes conveys immediately and powerfully what we mean by biodiversity and why we should care about losing it. (Darwin, naturally, wrote a book on orchids.) To an unusual degree for tropical plants, we know in detail where orchids can be found and where exactly we must act to save them. I applaud your interest in this book! It affords the clearest view of why we should care about the variety of life and what we need to know to ensure its future.

—Stuart Pimm, Ph.D.

Stuart Pimm, Ph.D. is a Doris Duke Chair of Conservation Ecology at the Nicholas School of the Environment and Earth Sciences at Duke University, in Durham, North Carolina. He is also an Extraordinary Professor of Conservation Ecology with the University of Pretoria in South Africa.

Delicate ionopsis (Ionopsis utricularioides) *is the prima ballerina of orchids; wrapped in ballet pink and shiny, diaphanous white, it is a favorite of poachers. Sadly, unlike most other wild orchids, it will grow in captivity.*

Duckweed blankets the deep slough surrounding a cypress tree island.

Wild Orchids in Southern Florida

The dense swamps, hammocks, and broad grasslands of southern Florida provide homes for nearly half of the native orchids to be found in Florida. Despite ongoing construction and development in these southern counties, the preservation of such areas as Corkscrew Swamp Sanctuary, Fakahatchee Strand Preserve State Park, Big Cypress National Preserve, Collier-Seminole State Park, Everglades National Park, and Big Pine Key assure the conservation of habitat for these precious botanical treasures.

Along with loss of habitat, overcollecting, periodic freezes, and other challenges have also contributed to the apparent loss of many orchid species, not just from south Florida, but from the United States as a whole. Local orchid enthusiasts, as well as professional botanists, are reluctant to declare these plants as permanently extirpated, but the reality is that some of them may never be seen again in the wild.

Species such as the rat-tail orchid *(Bulbophyllum pachyrachis)*, the small-flowered butterfly orchid *(Maxillaria parviflora)*, and Acuña's star orchid *(Epidendrum acunae)*—all discovered in the past half century deep within the Fakahatchee Strand Preserve State Park—have been determined lost. Arduous annual searches are still mounted in hopes of their rediscovery.

The history of orchid exploration in southern Florida goes back to the late 1800s and the work of A. H. Curtiss and A. P. Garber, who, between them, managed to find many of the mainstays of the southern Floridian orchid species. Such popular delights as the clamshell orchid *(Prosthechea cochleata* var. *triandra)* and the night-scented epidendrum *(Epidendrum nocturnum)* were first documented at that time. By the early 1900s, orchid hunting really began to come into its own. A. A. Eaton made a monumental trek, on behalf of botanist Oakes Ames from Harvard University, from Fort Myers on the west coast around the tip and through the Everglades and then north to Fort Lauderdale in what is now Broward County. This trip lasted several years and yielded not only new species for the United States, but also several species new to science. In some instances, it took nearly a century to recognize the significance of some of these early collections.

Although some of Eaton's collection data is imprecise, much of his work took place in what is now Fakahatchee Strand Preserve State Park. Fakahatchee looked very different when he was there, before the great logging and draining operations took place, than it does today. Ames would come to visit with Eaton and stay at the Naples Hotel near Gobbler Head (now the area of the Naples airport). They would take evening walks to see the ghost orchid *(Dendrophylax lindenii)* and many other specialties of the area that grew near the hotel at the time. Some of the areas that Eaton mentions in his travels include the Calusa River, Rattlesnake Hammock, Royal Palm Hammock, Flamingo, the various keys in Monroe County, and, of course, the Everglades in general. He spent much time in South Miami, Coconut Grove, Homestead, Black Point, and along the Miami River. Eaton also explored

Paul Martin Brown in the Fakahatchee Swamp looking at a night-scented epidendrum (Epidendrum nocturnum).

once-wild places, such as Brickell Hammock, that are part of metropolitan Miami today.

The original collections of three orchids destined to become new species to science have interesting histories. *Galeandra bicarinata*, the two-keeled galeandra, *Govenia floridana*, the Florida govenia, and *Spiranthes eatonii*, Eaton's ladies'-tresses, were all found some years ago, identified as other species, and then reassessed and described as new species more recently. In the case of both *Galeandra* and *Govenia*, plants were discovered for the first time in the United States in very small, local populations not far from Homestead. *Galeandra* was first found as a single plant in 1946 and identified as *G. beyrichii*, a Caribbean and South American species. Few plants were seen after that until recently, when a stable population of some size was found within Everglades National Park. In preparation for the treatment of *Galeandra* for the Flora of North America Project, Gustavo Romero, curator of the Orchid Herbarium of Oakes Ames at Harvard University, made the observation that the plants from Florida were most likely different from those from South America. I joined him in the field research, and subsequently we published it as *Galeandra bicarinata* Romero & P. M. Brown, a new species known to exist in Florida only in Everglades National Park and historically from western Cuba.

In 1957 Dr. Frank Craighead discovered a small population of a showy white *Govenia* in a dense hammock within Everglades National Park. Although not thoroughly convinced that this was *Govenia utriculata*, a species more common throughout the Caribbean and Central America, Craighead, Dan Ward from the University of Florida, and Carlyle Luer addressed that plant as such. Hurricanes over the next few years wreaked such havoc on the hammock that sheltered the *Govenia* that no extant plants could be found. Fortunately, both Craighead and Luer had taken numerous photos, and a few dried specimens of the plants were already in herbariums. In continuing my research for *Wild Orchids of Florida* I found it necessary to resolve this problem, and, after much comparison and consultation, I named the plants as *Govenia floridana* P. M. Brown, an endemic to southern Florida. Plants may or may not still be present in the wild.

The tale of *Spiranthes eatonii* is a very different one. During Eaton's south Florida collecting trip in 1904, he collected several specimens of a slender, white-flowered *Spiranthes* that bloomed in February in the dry pinelands. These plants came from what is now the Coconut Grove area. Upon his return, the pressed plants were shipped to Ames at Harvard University, mounted on herbarium sheets, and meticulously drawn by Ames. Originally identified as *Spiranthes lacera*, Ames then named them as var. *angustifolia*, for their narrow basal leaves. Shortly thereafter he crossed that out and wrote "new species. *S. eatonii*." For some reason, known only to Ames, the specimens were then put aside and not seen for many years. The species was never published.

In 1998, I was continuing my research on the orchids of Florida and came upon a different spring-flowering *Spiranthes* in northern Florida and adjacent Georgia. It soon became apparent that this unusual plant was not what others had thought it might be and that it was, at the very least, a new variety. I was preparing to name it

S. lacera var. *australis*, meaning southern. When visiting the Ames Herbarium at Harvard to search for additional specimens of this new *Spiranthes* (the species proved to range from eastern Texas to southeastern Virginia!), I was directed to a folder of old specimens in a bottom drawer that had not been looked at for many years. The result was the discovery of the Ames annotated Eaton collections, the same plants as the ones I was working with in northern Florida. Now, 100 years later, the orchid has an official name: *Spiranthes eatonii* Ames ex P. M. Brown.

This is certainly not a complete history of orchids in Florida. There are many more stories of discovery and lost species to relate, but these have been some of the highlights for me.

—Paul Martin Brown

Paul Martin Brown is a research associate at the University of Florida Herbarium in Gainesville, Florida. He is also the editor of North American Native Orchid Journal *and the author of* Wild Orchids of Florida *(2002);* Wild Orchids of North America *(2003); and* Wild Orchids of the Southeastern United States *(2004).*

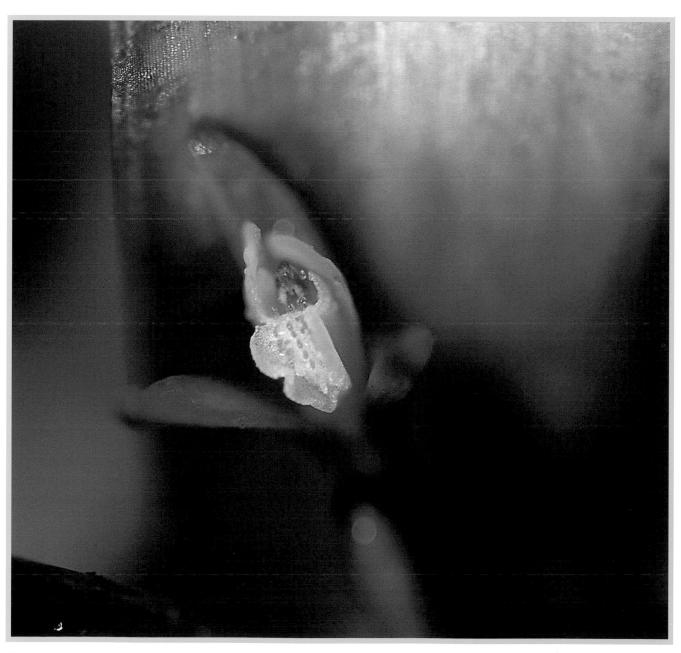

Gentian noddingcaps (Triphora gentianoides) *may or may not be native, but where it is found, it is abundant. It is also one of the most overlooked species in Florida, appearing primarily in gardens and bark mulch.*

*Uncommon in Florida and seen only in the southern counties, the clamshell orchid
(Prosthechea cochleata) blooms in all but the hottest months. Known in its several
varieties, throughout the West Indies, Mexico, and Central and South America, it
is a clamshell to some, an octopus to others, but to me, it's Queen Elizabeth I in
her high purple collar and pale green cape.*

Orchid Research, Economics, and Conservation

The orchid industry today is global, high tech, and growing at a rate far faster than the world economy. Orchid shows draw crowds that rival those at major sporting events. The discovery of showy new species continues at a rapid rate, and these plants are cultivated then hybridized to blend their unique genes with older sources. Computer-controlled greenhouses quickly produce, by seed and tissue culture, millions of plants that enter world trade and reach millions of retail buyers. World sales of orchid plants and cut flowers exceed the gross domestic product of many countries.

Despite this booming interest in orchids, human impacts continue to rapidly destroy the tropical forests where most of the 25,000-plus known orchid species grow. Our knowledge of the numerous microhabitats that host these species is poor, yet information about the narrow niches is vital to the success of orchid conservation and restoration efforts. Saving species means saving their habitats, which demands active management based on intensive scientific study, not a simply "hands-off" approach to preservation.

Most present-day "Save the Everglades" research is limited to a few subjects: exotic pest control, saving a few large animal species, and measuring the physics and chemistry of water flows, sea grasses, mangroves, and coral reefs.

However, our natural environments and the functioning of our global society form a single system that is vast and complex. The stability, dynamism, and robustness of this system depend on many factors that are sadly overlooked. Trillions of organisms including archaea, bacteria, actinomycetes, mycorrhizal fungi, wood decay fungi, yeasts, cyanobacteria (blue-green algae), green algae, and protists are informally lumped together as the microbes. We need to identify these microbes, learn their ecological roles, relate their activities to larger organisms (especially ourselves), and find a sustainable baseline pattern for preservation.

Orchid plants, their habitats, commercial orchid growing, the discovery and the introduction of new orchid species with commercially valuable genes, the long-term usefulness of soils, practical land management and restoration, our water supply, our social and psychological quality of life, and the world economy, are all interrelated. To save the orchids of Florida and to find more natural and yet economical means of ecosystem preservation, ecologists and microbiologists must explore and map the soil and water that is vital to the survival of epiphyte systems. This research will cost money, but ignoring the problems and continuing our reckless behavior will cost us, and the world, much more.

—John Beckner

John Beckner is the curator of the Orchid Identification Center at the Marie Selby Botanical Gardens in Sarasota, Florida.

All Creatures Here Below

". . . knowledge has not diminished by one iota the intense fascination that is aroused in all of us, who, having once seen [orchids] fall under their spell. They have a magic whose secret will always elude us, a blend of witchcraft and romance, that sets them apart, above all other plants—a sorcery that conjures up our deepest passions, from idolatrous worship through the pure spirituality of aesthetic veneration to fear, revulsion and mindless hatred."

—Luigi Berliocchi
The Orchid in Lore and Legend

Orchid books are too numerous to count, but none, so far as I have found, takes me where I want to go, into the realm of the spirit—on a ride, on a wild love affair with orchids and wilderness, across to an acquaintance with both the beauty and the wildness in my own heart. But, this book is not only about orchids. It's about us—our home, our land, the water that sustains us, and the wilderness within and without. Orchids are markers of our health and of our avarice. If the swamps are healthy, the water will be high when it should be, diminishing slowly over the dry season, seeping into our aquifers, cleansing itself on the way, and it will support orchids and all creatures here below. Orchids are enchanting bellwethers of the soundness of our environment.

Orchids are big business, too. In the United States, annual sales topped $100 million in 2002. Internationally, the trade in potted plants is estimated to be over $9 billion per year and the cut flower trade in orchids is at least $5 billion. In Florida, the ornamental plant industry is larger than either citrus or seafood, and orchids are a major part of that industry. In fact, commercial orchid growing is said to be the most profitable legal crop in the world.

These flowers are also economically important to the places where they grow naturally. Orchid ecotourism is booming, particularly in Costa Rica, Singapore, and other tropical nations. Orchids are a "significant indicator of natural conditions and a rallying-point for saving the environment," according to horticulturist John Beckner. He estimates that the needed natural services supplied by intact wetlands are worth somewhere near $33 trillion. For example, a swamp saved by the presence of rare orchids is worth nearly $20,000 per hectare per year in water supply purification and flood control, among other natural services.

Southern Florida boasts more than 40 species of native orchids, though some have not been seen for many years. Most are tropical species, their seeds brought from the Caribbean by wind or birds. They cannot tolerate freezing and have survived by staying in the deep swamps where the water protects them. If the water is high enough long enough through the coldest months—that is, if the swamps enjoy a normal hydroperiod—the water will absorb heat during the day and release it slowly during the night, insulating the tropical plants.

In the same way, too little water means insufficient insulation. A dry winter, or too many canals draining too much water from what should be swampland, combined with a sharp cold spell, could doom many if not most of our tropical beauties. At the other extreme, too much water backed up behind retaining ditches can drown them.

There are two types of orchids in Florida: epiphytic and terrestrial. Epiphytes, air plants, seem to live on air alone. They perch on tree limbs, anchored by aerial roots, and they collect water and derive their nourishment from photosynthesis like most other plants. Terrestrial orchids grow on the ground. A few cannot decide, so they perch on top of floating logs and thus maintain a distance of a few inches above the water at all times. Orchids display an incredible variety of colors, shapes, sizes, and strategies for survival. All seem to cast a magical spell.

But that spell does not ensnare everyone. While the major threat is alteration of the hydroperiod, more insidious and direct threats are posed by humans—habitat destruction on the one hand and theft on the other— for those who claim to love orchids are often the most destructive of them. Many plants are poached, but countless epiphytic orchids adorning trees throughout Florida are destroyed with the trees on which they rest as land is cleared for development. Occasionally, orchid lovers are called to rescue them before the bulldozers clear the land, and occasionally the plants are successfully reestablished in safe places. Terrestrial orchids adorning roadsides and byways suffer a different fate, as they are often unknowingly mown down as weeds.

In southwestern Florida, the many varieties of orchids live in prairies, mangrove swamps, hammocks, and sandy woods, but the most enigmatic are found deep in the strands and sloughs of our swamps. Like rainforests, these tropical swamps may be physically and psychologically threatening to some. Others see only the economic value of the swamp, and indeed, a generation ago, the great bald cypress gracing the southwestern Florida swamps succumbed to the snarl of the chain saw. To me, the sensual and spiritual riches of the swamps are enough. This deep, perhaps primeval, attachment to forests that most may know only through stories, books, and television, could—if it inspires us to action— be one of the swamp's great assets.

Travel with me through these wild places, seeking the essence of Florida's wild orchids and your own wild nature.

Like the swamp itself, the dingy-flowered star orchid (Epidendrum amphistomum) *is subtle. Both create fire in the hearts of some, but too often are dismissed, misunderstood.*

The aptly-named bee-swarm orchid (Cyrtopodium punctatum) *brings me to confront each flower as a bee would—huge, voluptuous, timeless.*

CHAPTER 1
History and Legend

"Early in the day we entered a narrow creek completely covered by branches of trees that interlaced overhead. . . . We saw more varieties of orchids than I have found in a single locality elsewhere, including specimens colorless and full of color, scentless and filled with odor that made the surrounding air heavy with their fragrance; some garbed somberly as a Quakeress, and others costumed to rival a Queen of Sheba."

—A. W. and Julian Dimock
Florida Enchantments

Lore and Legend

Orchis, of Greek legend, was a passionate youth. The son of a lovely nymph and a satyr, who bestowed on him a robust libido, Orchis let his desire overcome him. He tried to rape a priestess, but was condemned by the Fates who punish hubris. Wild beasts were summoned to tear him limb from limb, but, as poetic justice would have it, this beautiful, lusty young man metamorphosed into a slender, alluring plant with the organs that brought him to grief lodged in the ground. *Orchis*, which is also the Greek word for "testis", is still the name of the tuberous European terrestrial orchids, and the root of the word "orchid" and of Orchidaceae, the family of plants. (Luigi Berliocchi, *The Orchid in Lore and Legend*.)

In Europe, orchids were thought to be sexual elixirs, to stimulate lust, like an ancient Viagra. Medical books from the 15th century onward, and general herbal lore, offered elaborate recipes to release their aphrodisiacal qualities. European explorers brought back tropical plants from the remote corners of the world, and these orchids evoked intense emotions. Victorians had a particular passion for exotic, gaudy orchids.

This centuries-old fascination was not unique to Europe. Orchid legends throughout the world reflect sensuality, passion, flamboyance, an intersection with the spirit world, and, above all, awe. The Chinese succumbed to the allure of orchids and learned to cultivate them. In contrast to European varieties, Asian species were more commonly used for medicines and foodstuffs, religious and magical rites, and as an inspiration for art. In the Americas, Incas are thought to have used them in ceremonies, and the Aztecs cultivated vanilla orchids, one of the varieties we find in Florida swamps.

A happier folk tale from the American tropics suggests the importance the first Americans placed on orchids and offers an explanation for how orchids came to live in the crooks of trees. When the earth was young, the jungle's aggressive plants terrorized delicate blooms like orchids. These fragile orchids produced smaller and smaller seeds, until, at last, the seeds resembled tiny particles of dust. The Gods of the Wind took pity and blew the seeds into the protective arms of the great trees of the forest, where the orchids have lived happily ever after. I like that tale, and I like to think that seeds from the Caribbean were blown across the waters to grace the welcoming arms of Florida's cypress, oak, slash pine, pond apple, pop ash, and other trees and lianas of the swamps and hammocks.

Swamp History

Much of Florida, away from the glitz of the beaches, is very primitive—almost prehistoric. It is wild swampland that people, animals, and plants lived on long before the beach playgrounds, golf courses, and condos of the periphery brought millions to the state. The "real Florida" has a powerful life force that frightens and intimidates many but is a siren song to a few, a song of beauty and peace and serenity.

It is easy to detect the strands from above—dense, dark green fingers wandering through the yellow prairies. Names like Deep Lake Strand, East Hinson, Gator Hook, Roberts Lake, and Camp Keais evoke the history of these hostile lands, mostly named in recognition of the tough men and women who survived here or lost their lives before antibiotics and refrigeration and telephones and rescue helicopters.

However, most histories of southwest Florida are about conquering and taming—tales of battles won against Mother Nature. Men made a living by collecting snakes, frogs, and turtles, coons, pigs, and gator hides, or egret plumes for ladies' hats. Hunters sought deer and turkey for food, and shot the occasional Florida panther, well, just because. Few devoted much time or energy to conservation because the swamp's bounty seemed endless.

Others had a more devastating impact on the swamps. A major and probably irreversible blow to Florida's natural ecosystems occurred in 1928. By financing much of the Tamiami Trail —a road linking the Gulf of Mexico to the Atlantic —Barron Collier changed the way water flows through the entire southern everglades and brought commerce to the Gulf Shore. Developers likened their creation to the building of the Panama Canal. The so-called enemies were defeated—from snakes, panthers, and alligators, to mosquitoes, deerflies, sand fleas, and gnats. Wildfires were curbed. Books and articles lauded what was then called a victory, but the road itself is a death zone for wildlife, and the accompanying habitat destruction is immeasurable.

Dr. Frank C. Craighead saw this ruthless assault on nature not as progress but rather as a mark of shame. Craighead, an entomologist by profession, retired to south Florida in the 1950s. Since his first visit to the area in the 1920s, he had explored and written about the orchids and other flora of this delicate ecosystem. Craighead was ahead of his time. He saw that human survival in southwest Florida would depend on natural water flow and on the health of the natural environment. Though he was somewhat scorned back then, he now has his own special day of honor in Collier County. Craighead has passed on, but his legacy of science and conservation remains.

Phoenix Rising

Water flow is only part of the process. Fires in the prairies have been natural and necessary steps in the cycle for centuries. Usually lightning-ignited, but more often caused by human carelessness or evil of late, flames sweep across the over-grown prairies and parched ponds in the dry season, spring, setting the stage for explosive growth when the rains begin. Fire releases nutrients back into the soil. Now blessed with sunlight, orchids and other nutritious new foliage seem to burst from the earth like a phoenix rising from the ashes. New growth is easier for animals such as deer to digest; insects swarm on the succulent plants; and birds feast. The entire ecosystem is invigorated.

We humans plant ourselves and our possessions in what was once wilderness because we want to touch the edges of wildness, but in doing so, we thwart nature's plan. "Go around this house, fire," or "Skip this home," we seem to say, as if fire respected fence lines and No Trespassing signs. Global warming is a fact, and with it, hotter, drier seasons will, without doubt, bring more fires. As humans poke further and further into panther and wild-orchid territory, the fires will be termed "more savagely destructive" instead of necessary, fiercely beautiful, and rejuvenating.

Why the fascination with the apparently destructive force of wildfire? Is it the dread that we, too, might face destruction? Or wonder that we might be strengthened by adversity and rise from pain renewed? It is fire that separates us as humans, not tool use, as thought by those anthropologists who worked before Jane Goodall proposed her theories of tool use by chimpanzees. Fire uncontrolled, devouring all in its path and choosing its own trajectory, is guided only by its cousin, the wind, and reaches deep into humans' preverbal dread. Understanding and knowledge of what is to follow can transform that dread into hope for rebirth. Bred on fire, terrestrial orchids thrive. They hide below and then burst forth when enough light and water entice them into bloom. They are glorious!

In his song "American Pie," Don McLean sings "fire is the Devil's only friend." Not so. Fire cleanses hammocks and prairies of choking undergrowth and means renewal.

Swamp Tales

In Florida, America's orchid haven, many people have stories of the swamp and the orchids in them. I talked with some old-timers to try to learn how the wetlands ought to be by learning how they used to be. Oscar Preston "Longknife" Thompson, a fifth-generation southwest Floridian, was the son and grandson of hunting and fishing guides. His grandfather on the other side laid the first telegraph and telephone lines across south Florida. Oscar was a Floridian through and through—tall, lean, slow talking, and direct but shy. He started out his life gator hunting, and he appreciated the swamps and the animals and plants within them.

"I always liked the big cowhorn or cigar orchids," he noted. "There were quite a few, but a lot of those were poached because they were showy. They were on cypress trees and folks would cut down the trees to get the orchids."

Oscar loved the swamps and was always at home in them, but his eyes drooped when he compared the wildlands of his childhood to those of today. "I never dreamed that it would come to an end someday, but it has. It just gradually come to an end. The health of the swamps is picking up a little bit, but with all the pollution there is now, with all this farming and industry, I don't think it will get back to the levels of my youth."

Everglades photographic artist Clyde Butcher offered Oscar an opportunity to live in the swamp again and work as his assistant. The art and business mushroomed, and Oscar was able to display some of his own photographs on the walls of the Big Cypress Gallery. Tragically, before he reached the age of 60, Oscar passed away. Others who knew the swamps and prairies in the early days are much older. They will take their recollections with them unless we work fast to learn what our land was like before it was flooded with rapacious development.

John Beckner is still very much alive and with us. He has tossed me more information than I can possibly assimilate. As enthusiastic a person as I have ever met, John looks the part—horticulturalist's hands and fingernails, nondescript kneeled-on trousers, full beard, and intense, direct eyes. North of 70, Beckner combines the excitement and enthusiasm of a teenager with accumulated knowledge of a fiercely curious mind. He is dead certain about anything he is expounding on. His eyes twinkle with every new thought, and he forges through brush and swamp as if his legs could carry him well. He wants to launch a multidiscipline, detailed survey of all the Fakahatchee—"FlKahatchee" to him and other purists—to establish a scientific baseline for habitat preservation. He believes we can "use orchids as indicator species, flag-ships of what's going on. Epiphytic orchids are especially good indicators because they can be studied apart from soil conditions." He pauses for emphasis, brows raised, passions reignited.

Beckner's peer, Win Turner, is in perpetual motion—his body small and wiry, his character effervescent with high energy, his head covered with barely controlled thick, wavy, gray hair. He nails you with his blue eyes, and his brushy brows move like separate beings as he talks, gesturing with his never-still, working-man's hands. "How ya doing, sweetheart?" he asks gently, and gets away with it. I cannot help smiling.

He had first been introduced to me as one who knows wild orchids of Florida and their habitat as well as any, and I immediately knew I was in the presence of a gentle man who revered nature. Off we went through his garden, as elaborate as any botanical masterpiece, as he pointed to this plant and that, telling me where and how he got them—mostly from the swamps—and how he made each plant happy enough to thrive in captivity.

"There used to be hundreds of ghost orchids," he said, not afraid of hyperbole, but "all but a few died in the Christmas freeze of 1989." He was talking about one particular area of the swamp.

Looking for orchids on Mike Owen's famous "swamp walks." Humans are both the problem and the solution to restoration of native orchid habitat. Mike wants people to know and love the swamp and its orchids in order to save the natural ecosystem.

"They died because the water was drained too much and did not insulate the plants."

"Let's go check them out," I said.

So we did. He drove his Jeep, and after several other orchid stops and what could be a book full of recollections and instructions on caring for orchids, we came to an open gate on a dirt roadway with a clear No Trespassing sign. We could see the leveled woodlands beyond, now extensive agricultural fields. Though the gate was open, he was still for long minutes.

"Well," he finally said, "it's right in there," pointing to the patch of swamp on the other side of the gate.

We parked just on the legal side of the sign, walked in past it, and slid into what was left of the swamp. It was very deep—nearly too deep for me—up to my chest. I held my cameras above my head in a drybag, but if I stepped in a hole . . . I tried not to think about that. Win was dismayed. Too much water, the reverse of the normal problem, had literally drowned many of the orchids. We made our way inward, spotting a few tattered orchids along the way, but we were stopped by the water's depth. Win muttered. "This was a pond, but not too deep. These trees were loaded with orchids. We shoulda brought a canoe."

About 45 minutes into our journey, we heard a vehicle drive by, turn around, come back, and stop. "Hey! You in there!" It was a male voice. "It's dangerous in there. Do you know there are moccasins and gators in there?"

"Yes, sir," Win and I answered in unison. We did know, and while the gators were not of much concern—they are not used to humans and simply go away—water moccasins are another matter. If the snakes had come at us on the surface of the water, we would have been almost eye to eye—not good. I had been alert since getting in.

"Do you know you're trespassing?"

Gulp. "Yes, sir," we said again, in unison, with less enthusiasm.

"Then you come on out. Now."

We did, but it took nearly half an hour to make our way back to the edge of the road and climb out. "Hey, I know you," said the voice above the sheriff's deputy's boots. I was relieved to see an acquaintance when I looked up. No arrest. We shared our dismay at the effect of the backed-up water, flooding "good" swamp and cattle land in favor of new houses that should not have been built so far into the wetlands. The deputy, a man who knew and loved the wild world, just shrugged, rueful that his job was to protect the rights of those who were ravaging the space that should have belonged to us all. "But what can you do?" he said, dismayed. "It's private land."

"You was not the first one I almost drowned!" boasted Win, once we were back in the Jeep. He had been introducing newcomers to the wonders of south Florida for years. "Down at Collier Seminole there's a lake with lots of moccasins. I led a group. One Asian girl—very tiny, very pretty—started in swimming, but never complained. I respected that. I let her hold onto my arm and I walked her across, but that left me with only one hand to flick the moccasins away with my stick. I never said anything so as not to scare them. Everyone was jess looking at all the orchids and stuff. Never noticed."

He turned his enthusiasm to another tale, this one of a huge cigar orchid. "So big it would fill the back of a pickup. But it'll die. Ain't no pollinators around," he said sadly. The stories continued as we sped back to the concrete jungle. "There's orchids by the thousands in the buttonwood islands around Flamingo. I've even seen black panthers there. One stalked Steve Roberts and me, but we jess dropped the rabbit we'd killed and kept going." Scientists say there are no black panthers here, but what do they know compared to a man born to the swamps? Who's to say Win's understanding of Florida's backcountry—the prairies, glades, marshes, and hammocks—isn't gospel?

Crested coralroot (Hexalectris spicata). Nonphotosynthetic orchids are dependent on their fungal root associates. They live off sugar from a mycorrhizal fungus, which likewise stole the sugar from another plant—one that produces the sugar from photosynthesis.

Previous Page: Mounds of cloud and clear blue sky dwarf even the vast wet prairie of Big Cypress Basin.

The oblong-leaved vanilla orchid (Vanilla phaeantha) *is known only in Collier County. Gary Schmelz led biologist Mike Owen and me to an enormous vine hundreds of feet long, zigzagging across the tops of several trees in Fakahatchee Strand, that even Mike did not know existed. Early, we found several open blossoms, most of them high up. By the afternoon, they had wilted and closed, and the next day they were gone. Ephemeral, these large, delicate blooms highlight the fragility of life in the swamp.*

An individual inflorescence of the yellow helmet orchid (Polystachya concreta) can occasionally be isolated and the blooms examined.

It is said that all photographs are self-portraits of the artist's vision. If that is so, then I am a butterfly orchid (Encyclia tampensis)—*low maintenance, tidy, and hardy, neither big nor showy.*

Delicate ionopsis (Ionopsis utricularioides) *is often found suspended on tiny twigs over water, riding on its own inverted reflection.*

Giant ladies'-tresses (Spiranthes praecox) *is one of the most common terrestrial orchid species throughout the southeastern United States. Tall and spindly, it easily blends into the surrounding grasses.*

Most of wild coco's (Eulophia alta) relatives live in Africa and the tropics around the Caribbean. Large, unusually colorful, and showy for a terrestrial orchid, wild coco is Florida's only Eulophia.

This shadow-witch (Ponthieva racemosa) *and its buddies sprang up from the thick base roots of a royal palm plunked right in the middle of a swamp trail in Fakahatchee Strand. I was alone in the swamp—something I am told not to do, though I love being alone—and sitting in a few inches of water, maneuvering my equipment, working with the too-bright light, when I became aware of a presence. No, this time not the spirits of the forest, but as I turned to look, an equally surprised doe and I both startled. She wheeled and melted into the brush, leaving only her image in my eyes, and joy in my heart.*

Oncidium floridanum is Florida's only Oncidium. Win Turner says that at one time these huge, showy terrestrial orchids were collected throughout south Florida hardwood hammocks and sold in Miami for $50 apiece. Now if they are for sale it is more like $5,000. Normally, Oncidium is epiphytic. Normally, terrestrial orchids lack large pseudobulbs. Normally, terrestrial orchids are not so big and boisterous. Oncidium floridanum defies convention.

A coastal swamp orchid tolerant of salt spray, the Florida dollar orchid (Prosthechea boothiana var. erythronioides) has been collected from mangrove and buttonwood swamps along the south coast of Florida, leaving few natural colonies. Tropical, the dollar orchid does not tolerate cold.

Tiny frosted flower (Pleurothallis gelida) is truly a tropical flower, and the few remaining plants in the deep Fakahatchee sloughs are particularly sensitive to cold. Clinging to trunks and crotches of trees that are over insulating water in the dry winter months, the pale, diaphanous blooms are extremely difficult to find. I had to climb a tree to get near enough to photograph this orchid, and, under the macro lens, I saw a mosquito hawk poking all around the bloom. Is this the pollinator? Maybe.

The dwarf butterfly orchid (Prosthechea pygmaea) is very rare and endangered in Florida, found only in Collier County's deep swamps, though it is also known into Mexico and lands surrounding the Caribbean. This is the smallest of Florida's orchids, and finding one in bloom is like discovering the Littlest Angel dancing on the head of a pin.

Close encounters with ghost orchids (Dendrophylax lindenii), lips and pollinia beckoning transcend time and place. Confronting these mysterious orchids larger than life reduces the distance between human and flower and draws us into the spiritual interior of nature.

CHAPTER 2
Hooked on Orchids

"Good morning," the little prince responded politely, although when he turned around he saw nothing.
"I am right here," the voice said . . .
"Who are you?" asked the little prince, and added, "You are very pretty to look at."

—Antoine de Saint-Exupery
The Little Prince

Orchids capture the imagination. It is said they evoke an insatiable passion, a sensual appeal that can progress to possessiveness and then to obsession. Clearly, they can cast a powerful spell. But what changes admiration of this grippingly enigmatic family of plants to avarice, then to obsession? Where along the spectrum does love evolve into the quest for domination? Where does the hard face of greed appear?

Perhaps we must begin with the most basic of questions: What is the orchid's appeal? Harvard professor of botany, the legendary and aptly named Oakes Ames, writing in 1942, observed mildly, "Orchids for us are medicinal through the gracious charm of their flowers." Then he gets to the essence: "They are good for the soul."

How do otherwise reasonably normal people catch the fever? Naples Orchid Society president Tom Coffey, who is quite a normal human being, inelegantly, but succinctly, laughs, "Orchids must put out some sort of gas to make people go nuts."

❧

It was a nature photographer's perfect "Aha" moment. The ghost we came to see—ethereally pale, bizarrely shaped, leafless, nodding on a single long stalk to the almost imperceptible breeze—seemed to materialize in the gloom. It appeared theatrically lit to maximize the drama

of near iridescence in the half-light, dancing between the water below and the breeze-tossed leaves of the midcanopy. Was Billy Ghost flirting? Was he trying to entice some unseen lover who might, yet again this year, pollinate and thus enable him to perpetuate his own kind? Or was I being captivated, like so many other orchid lovers, by his beauty and elegance, his rarefied shapes and sensual patterns?

I gasped at the sight and forgot the cloud of mosquitoes, the bug spray dripping into my eyes, and the gouges in my shins from an hour's walk through roughly a meter of water concealing fallen logs, broken limestone, and lurking alligators. He was indeed awesome, this Billy Ghost orchid *(Dendrophylax lindenii)*—truly a star in the swamp's cacophony of images and deafening assault on the senses. And, as if to leave no doubt that he was in a league of his own, he boasted not one but two blossoms and two pods, indicating two blooms and pollination the previous year.

❧

I have read about orchid lust and have even met some orchid lovers who are a bit over the edge. I, too, was speechless the first time I saw a cigar orchid *(Cyrtopodium punctatum)* in full bloom—huge, golden, and red-orange, like a swarm of bees against the velvety gray cypress woods, still barren in February. It was sublime. But obsession? Not yet, anyway.

I am more intrigued by the detective element—by finding the shy wild orchids, catching them in bloom, hunting the pollinators, and uncovering the myriad strategies these highly evolved plants have devised to grow and proliferate, and how they trick insects, birds, fungi, and trees into serving them. Add humans to that list. Am I anthropomorphizing? Absolutely, but I am in good company. Growers do it all the time, and so does Mike Owen, biologist at Fakahatchee Strand Preserve State Park. He named Billy Ghost for Billy Snyder, who found it in 1997. There is a Clyde Ghost, named for famed photographer Clyde Butcher, and, I am extremely proud to say, as of June 30, 2002, a Connie Ghost. Its first-ever bloom fully opened on my birthday, and I photographed the 24 hours it took to open from bud to bloom. Mike had been watching this plant for seven years, and he alerted me to the spike and bud, but it was my daughters and I who saw her open that special day.

Mike is also the one who led me to Billy Ghost's secret location. Mike's bailiwick, Fakahatchee, is the center of what, with over 40 native species, is known as the orchid capital of the United States. This royal palm, bald cypress, and epiphyte jungle, bordered by marsh prairies, hardwood hammocks, and cypress domes to the east, north, and west, and salt marshes and mangrove forests to the south, has the largest concentration and variety of orchids in North America. The Big Cypress Basin, next door to the east, runs a close second.

Fakahatchee is like a wide, slow-moving river, 22 miles long and between one and five miles wide, flowing south through southwest Florida to the 10,000 Islands of Florida Bay. Its cypress- and palm-wooded strands are shallow and may dry out in the dry season, but its sloughs, pronounced "slooz," are deeper, marked by pop ash and pond apple. Sloughs do not dry out.

Some orchid species hide in the most remote sloughs. Others, such as the butterfly orchid *(Encyclia tampensis)*, thrive near humans and can be found on trees in gardens, parks, waterways, and on roadsides all over southwestern Florida. In fact, one particular tree hosts dozens of plants not 50 feet from one of the busiest intersections in Collier County, which is among the fastest-growing counties in the nation. However, just finding the rarest plants, then finding them in bloom, is a full-time job and requires a lot of luck. An informal army of researchers and recreationists is on the lookout for orchids all the time. When someone alerts me to a bloom, I follow directions and retrace steps to locate the plant. Even then it is not easy, as some orchids bloom for a day or less, and most are nearly impossible to find in the swamp.

Wild Orchid Fever

Paul Martin Brown, whose recent book *Wild Orchids of Florida* is the guide of choice among orchid enthusiasts, called me one day in May, I think to test me. He told me that one of three species of orchid lacking all traces of chlorophyll (mycotrophic) was in bloom near Fort Pierce, a four-and-a-half-hour drive away. This orchid, crested coralroot *(Hexalectris spicata)*, is entirely fed by its mycorrhiza in a symbiotic relationship between its roots and a specialized fungus. The

Epiphytic orchids on a limb in Mantadia National Park in the eastern rainforest of Madagascar.

next morning I was on the road by 6 a.m. It took a frustrating hour to find the plant, a 20-inch terrestrial orchid which, from a distance, looks like a dead stick. Up close I could appreciate its subtle tones of muted pink and soft brown. It took another hour to photograph the plant. Then I drove home. The trip was 11 hours door-to-door. Is this Orchid Fever?

My first wild orchid was introduced to me by my daughter, Lea Borkenhagen, in Borneo. It was a daringly red, trumpet-shaped, ground orchid, bold against the half-light of the brown rainforest floor. Then there was another in northern Madagascar, a white terrestrial *(Phaius humblotii)*, diaphanous in a tropical downpour, calling me to its space under a low limb. There were others in Madagascar—pink terrestrials and white epiphytes—hanging from limbs all over the remote eastern rainforest where I was working. There were orchids in Sulawesi and Komodo, Bali and Peru. Most were nameless. All were adored for a moment then left to be protected by the remote terrain that surrounded them.

Despite these early encounters with untamed beauties, I blame Larry Richardson for first infecting me with wild orchid fever. Nature photographer and biologist at the Florida Panther National Wildlife Refuge, which comprises the top quarter of the Fakahatchee Strand, Larry exposed me to the swamp's many wonders years ago. Unlike the state park to the south, the panther refuge is closed to the public; thus, it is easier to find orchids in profusion in isolated pockets, despite poaching, the effects of water deprivation, and the reduction or absence of pollinators.

While Larry sparked my love for orchids and their strange and seductive lives, I hold Mike Owen responsible for feeding that passion. In May, 2002, Mike took me scouting in Fakahatchee Strand to check for spikes signaling incipient ghost orchid blooms. By this time, I had learned to find the well-camouflaged plants—leafless epiphytes with a star of gray-green roots bearing short white markings and chlorophyll-green root tips. As the main trail branched away from our goal, we saw the first of numerous freshly tied pink markers. Unlike the flagging of seasoned woods people, these ribbons were too long, spaced much too close together, and clearly meant for someone who was not swamp savvy. We pulled them off as we went, increasingly alarmed. Then we saw the unmistakable mark of an orchid thief. A low-lying ghost orchid known to both Mike and me had been scraped away from its pop ash branch along with accompanying fungi and ferns. To this day, Mike has not found the culprit.

Hundreds of species of orchids like these, native to the montane forests of Minahasa, Sulawesi, grace the thousands of islands of Indonesia.

Susan Orlean's best seller, *The Orchid Thief*, set in and around Fakahatchee, and the movie version, *Adaptation.*, chronicle this obsession by following the trail of one convicted thief and his network of weird cronies in south Florida. Eric Hansen's best seller, *Orchid Fever*, weaves a similarly complex web of love, lust, and lunacy from continent to continent. Some will stop at nothing to see and own the rarest, most beautiful wild orchids—to name them, tame them, pluck them from the wild, possess and control them. The ghost orchid is at or near the top of that list.

Feeding the Passion

Far from stemming the passion for these flowers, the Native Orchid Restoration Project (NORP) hopes to feed it. Not only poaching, but also loss of wetland habitat, reduced hydroperiod, and loss of pollinators, all directly attributable to the relentless push of humans into formerly pristine swamp, have reduced orchid numbers drastically. NORP was born when a small group from the public and private sectors came together to create a master plan for orchid restoration. Weaving science, public relations, historical analysis, and missionary zeal, they envision an

This ghost orchid (Dendrophylax lindenii) *bud was photographed by time-lapse interval camera, but the plant itself aborted blooming because a bug bit into the bud.*

international network of people working to restore wild orchids to their former abundance.

To save and restore wild orchids, many more people have to see and know them, which is impossible if the orchids are hiding in the deepest swamps. Enter Caribbean Gardens, a popular privately owned zoo and tropical garden in Naples, Florida, open to the public but locked at night to deter theft. Director of Education Tim Tetzlaff is "excited about being able to display these rare orchids in order to educate our guests about the wonders of southern Florida . . . and what they can do to help protect some of its diversity."

I have learned more about wild orchids than I ever would have imagined, and I keep hunting to learn more. In the spring of 2002, Mike Owen and I walked way into the deep swamp and found an enormous vanilla vine *(Vanilla phaeantha)*, hundreds of rigid *(Epidendrum rigidum)*, night-scented *(Epidendrum nocturnum)*, dingy *(Epidendrum amphistomum)*, clamshell *(Prosthechea cochleata* v. *triandra)*, ribbon *(Campylocentrum pachyrrhizum)*, and elephant-ear *(Liparis elata)* orchids as well as countless endemic bromeliads and ferns. We had each discovered a new ghost orchid plant on this outing, but Mike, leaning a raised elbow against a tree, challenged me to find more.

"Do you mean like this?" I pointed to a large, healthy plant with a small spike and tight bud right under his arm.

The piece de resistance, however, was Mike's revealing to me what is now Connie Ghost. This plant is particularly vulnerable because it is close in, though protected by a deep slough and a 12-foot resident alligator. I set up my camera and documented the blossom's unfolding with time-lapse photography. I know for a fact it takes 24 hours to open—new scientific information. I also set up equipment to photograph the pollinator, but it never came. One secret is all Connie Ghost will reveal for now.

When will she bloom again? Will we be able to catch the moth in the act of pollination if she does bloom? I hope so. For now, I am pleased to be considered trustworthy enough to know of this apparition. I will guard her secret.

Spreading the Fever

I have taken many first timers with me into the home of wild orchids and their kin. Most were prepared and positive despite the pesky mosquitoes and slippery footing. An adventure with Scott Stewart was one of the most bizarre and fun. I had set up a camera to take a photograph every 45 minutes as a ghost orchid bud emerged and had to hike to the site to change the film and batteries every day for a week. It was a 30-minute straight shot along a tram, then along a shallow slough and into a small hammock of oaks and cypress. Scott, though just out of undergraduate school, had already established himself as a brilliant student of orchid-fungus associations, but he had worked primarily on terrestrial orchids in temperate climates. He was excited to come along and did not complain even though the mosquitoes were getting the better of him. As we rounded a thicket and the hooded camera and tripod we'd set up the day before were visible, he darted in front of me to see the ghost orchid bud.

"Ahh!" turned to "ohhh dear" as we both saw that the nearly mature bud, yesterday full and ripe, now had a big insect bite right on top. That would abort the flowering for this year, though I wanted to keep the camera up one more day just to be sure.

Scott started rushing around the hammock, finding literally hundreds of epiphytic orchids, his thrill for the hunt building in the process. "*Rigidum*! And it seems to be in association with *Nocturnum* nearly everywhere.

Do they share the same fungus? . . .This moss and fungus seems likely, but why no orchids?" He was not so much talking to me as talking to himself, in another world. But, like a puppy, he checked frequently to see where I was, for I knew the way out of the swamp, and he was all turned around. The secret to orienting in the Fakahatchee and Big Cypress lies in the water. It always flows south.

"Look at these mushrooms, Scott," I called. "Could they be significant?" They were everywhere, fungi of a different sort than we'd seen earlier. He took scrapes from various places on the underlying limb, placed them in ziplock bags, and marked each one—ever nimble of mind, but also open to the pure joy of being in the presence of the abundance of life and the poetry of its beauty. Orchid fever is highly contagious and nearly impossible to cure.

❦

Denouement: Billy Ghost died the winter of 2002. Maybe he bloomed too much. Maybe he was old. Maybe producing pods several years in a row exhausted him. We can only hope that his progeny are many and thriving throughout the slough. We won't know for years—if ever.

Mushrooms thrive on a decaying log.

Epiphytic butterfly orchids (Encyclia tampensis) *are abundant in south Florida as well as in Cuba and the Bahamas.*

Previous Page: *Alligator flag fringes the deeper waters of the slough in bright green.*

Some clamshell orchids (Prosthechea cochleata *var.* triandra) *can be giants—like this Henry VIII of clamshells—with spikes several feet high, but most are more diminutive aristocrats.*

The shadow-witch (Ponthieva racemosa) *seeks damp, shaded places in the swamps and woodlands.*

Scott Stewart, Lee Hoffman, and Mike Owen return a confiscated stolen night-scented orchid (Epidendrum nocturnum) *to a remote area of the Fakahatchee Strand Preserve State Park. A thief had taken a chainsaw to the cabbage palm on which this orchid clung, cutting the entire top of the tree off to get to the orchid. He was caught with the orchid as he was leaving the park. The plant was still thriving a year after it was replaced.*

It is rare to find a fully open night-scented orchid (Epidendrum nocturnum), *for most are self-pollinating.*

The leafless ribbon orchid (Campylocentrum pachyrrhizum) *is known only to live in the Fakahatchee Strand. Though the individual flowers do not grab you as you walk by, the totality of the root system, embracing trunks of trees many feet up, is impressive. Usually found deep in the swamp, one well-known stand on two adjacent royal palms has recently been raided by poachers.*

Epidendrum rigidum *likes to share its perches with its larger kin,* Epidendrum nocturnum. *Named for the rigid central keel on its leaves, it is hard to notice the tiny, green-to-rust flowers on this enigmatic orchid. Hardy,* E. rigidum *is one of the most common orchids in the swamps.*

Author and orchid expert Paul Martin Brown called me to tell me where the crested coralroots (Hexalectris spicata) were flowering, a four-and-a-half-hour drive from my home. Of course, I went the next day, and, despite his explicit directions, I had a hard time finding the plants. Their pastel colors blended into the leaf litter in the underbrush of the oak hammock.

Though native to Mexico and Central America, commercial vanilla plants (Vanilla planifolia) escaped and have become well established in Dade and Collier Counties. Vanilla orchid vines grow readily from cuttings, and their rich, butter yellow hue is popular with horticulturists. The large flowers are delicate and short-lived. They open early in the morning and are gone by the heat of the day.

The elephant-ear orchid (Liparis elata) *cannot seem to decide if it wants to be a terrestrial orchid or an epiphytic one—though it is classed as a terrestrial. It perches elegantly on floating logs in the open sloughs. Photographers and anyone wanting a close look must get quite wet.*

Ghost orchid (Dendrophylax lindenii) *detail—a ballerina's gossamer scarf trailing, dancing, in the caress of the breeze.*

The toothed orchid (Habenaria odontopetala) *smells bad and looks just plain messy—until you look closely at each individual, elaborately delicate flower.*

*Early morning light
in the slough.*

*Thoroughly naturalized, the African spotted orchid (Oeceoclades maculata) has spread like a
weed since its appearance in Florida some 30 years ago. Its original pollinator may have been
left in some African savannah, but in Florida it makes do with rain. It is one of the few orchids
where raindrops dislodge the pollinia and allow it to fall, pollinating itself. That frustrates orchid
lovers, as the flowers do not fully open unless there is no pollinating rain.*

The terrestrial orchid (Habenaria distans) is, according to Paul Martin Brown, one of the rarest orchids in the United States. Found only in Collier County, its siblings inhabit Belize and other Caribbean locales.

Wild coco (Eulophia alta). People looking for a link to their natural environment fail if they do not give nature time to be seen, experienced. She reveals her secrets slowly but only to those with knowledge, patience, and a heart ready to receive her gifts.

The white butterfly orchid (Encyclia tampensis *forma* albolabia) *is rare. The petals and sepals are pale green, and the lip is white.*

CHAPTER 3

Love and Domination: Who Captivates Whom?

"From the myriad of shapes that cloak these plants come flowers of such forms and colors, evoking a fantasy of psychedelic imagery—from animistic theology to the most exquisite mimicry. Magnificent and perfect-designed, decorated and embellished by invisible hands—and then, by the most divine irony, orchids have been hidden from human kind for millennia in the highest branches of the great trees of these long inaccessible tropical forests."

—*Luigi Berliocchi,* The Orchid in Lore and Legend

A desert-bred girl, I dreamed of exuberant greens, mosses and ferns, silent waters, dark places among overhanging limbs, and peace. *Green Mansions*, by W. H. Hudson, was one of my favorite books. In my imagination, for I had never seen rainforests or swamps, the Everglades were nature's greenhouse, alluringly cool and damp, primordial, a trip back to the beginning of time and innocence. I later learned that I was right . . .

Within seconds I was intensely happy. It was the first time I had waded into the cool, dark waters, but I was immediately at home. While others were exclaiming "ugh" and "yuck," I was quiet, exulting in the same peace and serenity I experience when entering Notre Dame. The music was birds and insects, frogs and the breeze on the palm fronds. The stained glass was all around—emerald facets dancing on all sides, inverted on the water's surface, playing tricks with my eyes and my heart. Tree twins mirrored each other, one reaching high into pieces of blue, its reflection plunging deep toward the center of the earth. Life permeated every crevice. The sheer wonder drew me deeper. My human companions ceased to exist,

eclipsed by the ferns and orchids and spiky bromeliads, the unseen creatures, the birds above, the profusion of life. After a year of "doing" Naples, I had found the authenticity I craved, and it was only a few minutes away.

Just a few days passed before I was back, early in the morning—photographers' time. Mist was rising off the exposed water and huddling in the shade of trees and shrubs while lianas and ferns were seeking the first light. I was but five minutes from my car, but thousands of years back in time. Braids of thorny smilax vines grabbed at my trousers. Lilies and bladder-worts warned of deeper water, and alligator flag whispered, *here it's even deeper*.

Oddly, I found this swamp haven only after traveling to the other side of the world to visit African and Asian rainforests. The first time I looked directly into the eyes of a wild chimpanzee, both of us squatting on the forest floor not three feet apart, I knew nothing could ever dislodge the serenity he disclosed in my heart. It has been shaken at times, yes, but remains strong. Here in our own swamps, I can dust it off and wear it comfortably.

Love Gone Wrong

For three million years humans lived as any other creature, at the mercy of the forces of nature, without any control over their environment. Along with the harnessing of fire and the development of consistent food sources—domesticated animals and agriculture—humans moved inexorably toward domination of the natural world. One of the oldest written edicts divines that the world was made for man, and man was made to conquer and control, to "have dominion over . . . every living thing" (Genesis 1:28).

It would seem easy to accept that we should not destroy the planet that supports us, but in the 21st century the Judeo-Christian community is not unanimous on its interpretation of Genesis. Duke University graduate student Kyle van Houtan and Dr. Stuart Pimm, Doris Duke Professor of Conservation Ecology at Duke's Nicholas School of the Environment, define the array of thought on the subject. The spectrum goes from a concept of "earthkeeping," led by Greek Orthodox Patriarch Bartholomew I, which recognizes that there is a biodiversity crisis and embraces it as "an ethical issue of great concern," through those who are skeptical about an extinction crisis, to groups van Houtan calls "indifferent." The latter self-identify as having a "pro-family agenda" and do not address the issue of endangered species or extinction at all.

In our continued march toward control and domination, we decide which plants are weeds—defined as not useful to humans—and which can be captured or cultivated to serve our purposes. Many terrestrial orchids, mown down along road-ways, withered by insecticides, or turned under with plows, have fallen into the first category, while secretive epiphytes become the victims of greed and obsession. How and why do forces compelling us to control our environment some-times metastasize into the drive to possess the rarest of wild orchids at any cost?

Some We Know

Perfectly suited for life in the damp, bald cypress has wood that is durable and rot resistant. A thousand years of nature's work was felled in little more than a decade—cut, left to dry, then dragged out of the Big Cypress Basin to form the hulls of 1940s Navy minesweepers and later to build boat decks and new houses. Only two stands remain of the East Coast's equivalent of the mighty redwoods—one at Corkscrew Swamp Sanctuary and the other at Big Cypress Bend at the bottom end of Fakahatchee Strand. Odd singles remain, too, usually saved by the embrace of strangler figs deemed ruinous by loggers. Bald cypress are being reestablished in the strands while leaner pond cypress still abound in hammock domes and edges. The hatrack, or dwarf cypress, natives of the wet prairies, are too gnarled and diminutive for our rapacious tastes. They remain bizarre elfin reminders of their vanished cousins.

The branches of the smaller, young bald cypress now serve as the forest's canopy, with pond apple, pop ash, and maple as midcanopy shade. Native royal palms shoot high above, but offer little cover, while stockier sabal, also known as cabbage palm, offer too much shade. Below is the limestone substrate risen from the sea. Erosion and acids—carbonic acid of dissolved carbon dioxide and humic acids of plant decomposition—are slowly eating away at the exposed stone. On top, needles from cypress, one of only two deciduous conifers in the U.S., and weedy-looking slash pine, the only pine to inhabit these wetlands, accumulate along with leaves from other trees and shrubs, including willows. Everything decomposes to create and build peat. If it does not burn, red maple and oak establish themselves, and voila! Hundreds of years later, there are hammocks.

The trams, former rail beds where, close to 60 years ago, tracks were laid to haul the ancient cypress out of the swamp, serve as "quickie" hammocks. The tracks are all gone, along with the big trees, but the raised tramways remain and host species that need a drier habitat. Many terrestrial and epiphytic orchids love these humid, protected hammocks, and royal palm, gumbo limbo (also known as Englishman's tree because its bark is red and peeling), paurotis palm, dahoon holly, and coral bean all jockey for sunlight in these higher areas. Other plants, particularly epiphytes, seek the protection of the deep water, where they hang on limbs up to the limits of the water's insulation—10 to 12 feet high—where light reaches but temperature extremes do not.

The Big Cypress Basin has the highest densities of rare plants—ferns, bromeliads, peperomias, and orchids—in the continental U.S. Much of the plant life of this wilderness is as uncatalogued as it is in the remote rainforest. We regularly see varieties of fungus, ferns, and cycads too numerous to count, and spot fresh-water sponges encircling submerged branches, usually cypress.

Twenty-five years ago we came perilously close to losing much of this wilderness to development. The private land that is now Big Cypress National Preserve, all 729,000 acres of it, was destined to become a massive airport complex, pitting the Collier financial juggernaut against environmental concerns. Florida had a tragic record of wetland destruction with little interference from the governmental units charged with protection. Yet this time, the state and federal government came through. The airport scheme was stopped, and Big Cypress was purchased for the public. Though some private holdings still pockmark the inside, Big Cypress has, for the most part, been returned to all creatures here below.

Mike Owen in Fakahatchee swamp.

Ghost Stories and Swamp Walks with a Dead Man

Skinny and painfully steeped in the teen years in appearance, but sharp and self-assured in demeanor, Jake Heaton had managed to talk his father into buying a tent and driving him the length of Florida to join our survey of the inner slough. For three days, a select but ragtag group of hard-core swamp enthusiasts—10 of us including a fern specialist; an expert with encyclopedic knowledge of bromeliads; orchid, reptile, and tree specialists; and me, the photographer—trudged into the unsurveyed interior of Fakahatchee Strand Preserve State Park with Mike Owen. Mike was featured in *The Orchid Thief* and the movie based on that book, *Adaptation*. Though his character was killed off in the film, I assure you that the real Mike Owen is very much alive and uncommonly lively.

"You are entering America's Amazon jungle," shouted Mike, to his motley band, though we all knew it well, "Florida's Grand Canyon." The slight dip in otherwise flat land draws the water south. The wooded part is called a strand, and the slightly deeper part—we are talking waist deep instead of thigh deep—is a slough. But jungly it is, and as much of a tropical rainforest as we have in the U.S. Really, it is part of the Caribbean ecosystem that got stuck onto Florida.

Cyril was there, a wiry elf who is ageless (but well, well past middle age). He spun some riveting tales about his youth as an itinerate salesman in pre-independence Africa. Russ got time off from policing in Miami to come. John, as handsome as he was in his New York television days, and Nancy, in her signature preppy shorts, were there as well. We paid little attention to the heat, the insects, smilax thorns, submerged rocks, cypress knees, mud, and lurking gators. Instead, we shared our enthusiasm for wild orchids.

It was a constant contest to spot and identify orchids. Jake was clearly in the lead, both in number and rarity of his findings. "*Purpurea*!" he shouted while we were still on the tram, and we all rushed to see. "Pink and white morphs. Don't step on this one with pods." Riding five miles and walking another five, we pitched our tents on the dry pieces of the tram and headed into the slough, heavy and dark and three to four feet deep in clear, tannin-colored, slowly moving fresh water.

"*Ionopsis*!" sang out Jake. We all slushed over to the diminutive, diaphanous blooms clinging to a tiny twig. "Clamshell in bloom here!" Mike lobbed back, "and ripe *nocturnum* pods on the next tree." "*Rigidum*!" until we were simply counting orchids, some in bloom, most not. We found many ghosts,

but none in bloom—too early in the year. It was a challenge to spot the leafless orchid root stars clinging to pop ash or pond apple, cypress or oak. "If you don't find enough," Mike teased me, leaning his arm against a fair-sized cypress wound around with a thick arthritis vine, "we'll leave you out here."

So it went, finding, identifying, counting, describing, all in good spirits despite the horrific heat, the humidity, and the biggest cottonmouth water moccasin I have ever seen. Threatened by our presence, the snake coiled with its head up and its mouth open, challenging anyone to approach. No one did. Nor did we pose a threat to the hundreds of orchids we found that weekend. Their very remoteness guaranteed them some safety, and all of us had been carefully vetted, over a number of years, by Mike. We were not orchid thieves.

An orchid lover surveys the deep sloughs of Fakahatchee Strand.

Swamp Love

Sitting at an orchid show I joined a conversation between two old pals and rivals. Each was bragging, as old men sometimes do, of the hundreds—nay, thousands—of orchids they had collected in their youth, verbally leading me along canals and mangrove swamps, through thickets and ponds. These men connected with something wild, with wildness, and they saw that I did, too.

"Them orchids in the Fikahatchee now," one said, pronouncing it the way the real old-timers do "ain't nothin' compared to what they used to be." He waxed poetic about the bags and bags of orchids he used to bring out of the Fakahatchee and his main haunt, Homestead, to sell.

"They's all gone now since they put the road in, and that canal," said the scruffier one. "Now it's all fulla melaleuca." Draining had begun in earnest in the 1920s, and the Australian melaleuca trees were introduced, allegedly to suck up excess water. Native to Australia, these trees have no natural predators in Florida and have aggressively pushed through and destroyed much native vegetation. Florida's endemic plants don't grow in melaleuca stands. Good-bye orchids.

Both men are now successful commercial orchid growers. I asked if having greenhouses was the same as collecting. "I cain't push through those sloughs anymore," mused the taller one, avoiding the question, "but sometimes I gotta go anyway." The tough macho veneer was gone. He was wistful, his psyche fully connected to what was lost.

For these men, the need seemed to be engagement with the natural world of wooded swamps and open prairies rather than the need to control the things that excited their spirits. Rough-edged though they seemed, they understood and appreciated the poetry in wildness and the untamed chaos of the swamps.

Fear and Disconnect

"Aren't you afraid to go out there?" Not at all. Cautious, yes, but afraid, no. In the jungle, my heart sings, my whole body relaxes, I am in harmony with myself. The Florida swamps and forests are seductive, like all rainforests I have experienced.

"What's the most dangerous experience you've had out there?" Getting in my car and driving the 25 miles to the swamps. Or maybe wondering which of the hundreds of biting mosquitoes might carry disease.

"You walk in that water? Ugh!" It's cool and clear and slowly draining south—truly a shallow sheet of water. They say you can drink it, but I have never tried.

"Aren't you afraid the snakes and alligators will get you?" It's the gators on the greens that are the most dangerous, those used to human presence. In the swamp, they are shy. They stay out of my way most of the time. As for snakes, yes, I occasionally see them and stay clear.

The real question, for me, is why the fixation on fear, on the possible terrors of our human origins? Is the profound disconnect from the natural world due to lack of knowledge and understanding, or does it stem from the fact that out there we are not in control? Seeing one wild ghost orchid, one delicate *ionopsis*, one profusion of clamshell blooms, or just one bee-swarm orchid would surely help us reconnect with our souls, our vital centers, and to the rest of the planet. Orchids, untamed, are found on every continent humans inhabit. Through wild orchids—not the beauty-contest cultivated sort—we all might recognize our common humanity, our human roots, our vulnerability, and our frailty.

This night-scented orchid (Epidendrum nocturnum) *pod is* dehiscing, *or releasing, hundreds of thousands of dustlike seeds to the wind.*

The night-scented orchid (Epidendrum nocturnum) is usually self-pollinating. Before it fully opens, its pollinia-bearing anther bends toward the stigma and releases the pollinia to fertilize the plant.

The yellow cowhorn orchid (Cyrtopodium polyphyllum) *is just as content to brighten a vacant lot as it is to populate the pinelands of Dade County.*

Native Orchid Restoration Project (NORP) member, Tom Coffey, nails butterfly orchid shoots to trees at a local golf course so golfers will learn to appreciate this hardy native species.

Butterfly orchids (Encyclia tampensis) may flower any time of year but are most prolific in June and July, the beginning of the rainy season. This huge specimen luxuriates on a horizontal cypress log just a foot or two above the water. It was lightning, thundering, and beginning to rain as I sank into the water on my knees to photograph this showy plant. I saw many of its progeny on surrounding trees.

Without human help, the cigar orchid would be relegated to a tenuous, unpollinated existence, eventually disappearing from earth forever. Enter biologist Larry Richardson, who takes tiny forceps to pluck the pollinia, a single waxy cluster of pollen half the size of a pinhead, and implant it in the pistil of another plant in the hope of sustaining the species.

Once common in southern Florida, remaining cigar orchid (Cyrtopodium punctatum) plants are scattered and highly endangered. Pollination occurs only between different plants, and the bees that formerly performed this service by way of pseudocopulation are now absent.

Most clamshell orchid (Prosthechea cochleata var. triandra) flowers are diminutive, and with their high, royal purple and gold collars framing the central column above five trailing, pale yellow sepals and petals, they are nothing less than elegant royalty.

Found only in south Florida, pine-pinks (Bletia purpurea) love wet pinelands and wooded areas of the swamps. Bright and open to the breeze, they stand out well from the drab prairies and hammocks where they thrive.

Mike Owen collects seeds from a ripe pod of the dingy-flowered star orchid (Epidendrum amphistomum).

The tiny dingy-flowered star orchid is often overlooked, like a watercolor dream nearly overpowered by its loud swamp neighbors.

Delicate ionopsis (Ionopsis utricularioides) *is a twig epiphyte producing delicate, diaphanous, pink and white flowers loved by collectors. Flowers are widespread but sparse and hard to see because of their delicate color and diminutive size.*

From the top, it is easy to see the spiral formation of the flowers of the terrestrial giant ladies'-tresses (Spiranthes praecox).

The best reward of photographing Epidendrum strobiliferum *is the "oohs" and "cools" I get from the real hardcore orchidophiles. Few have actually seen the plant. The French have a phrase describing my fascination with this orchid:* Joile laide. *It means "not conventionally pretty but still alluring."*

The lawn orchid (Zeuxine strateumatica) is a naturalized Floridian. Popping up like miniature crocus in lawns, gardens, borders, and even in the deep woods, Zeuxine is an illegal but benign alien that probably hid in grass or other seeds from Asia or Africa. Whatever its heritage, it is a welcome addition to Florida's wild orchid scene.

The leafless beaked orchid (Sacoila lanceolata var. lanceolata) is one of our most striking and most common terrestrial orchids, yet it is all too often overlooked or mown down along roadsides or in medians. Usually coral, other morphs are also common—white to pale pink, brick red to greenish, or even golden brown. When in flower, they are leafless.

A tall weedy terrestrial, Malaxis spicata *seeks sunlight but is firmly based on damp soils. Like* Liparis elata, *it often grows on floating logs.*

Commercial vanilla plants (Vanilla planifolia). Vines with large, well spaced leaves, readily climb trees and other structures. The best vanilla now is produced in Madagascar from vines originating in Mexico. As Madagascar lacks the pollinators, commercial vanilla growers hand pollinate each flower to produce the pods from which vanilla flavoring is extracted.

CHAPTER 4

Lives of Deception

"Ephemeral, blooming only to be pollinated and reproduce, the orchid, some say, is the perfect flower, both male and female in form, shaped to perfection by an unseen hand, sensual, visually and sometimes olfactorily exciting."

— *Connie Bransilver and Larry W. Richardson*
Florida's Unsung Wilderness: The Swamps

Where Spirits Dwell

Early Spring. The cool water was still deep at McBride's pond—up to my armpits in many places, and deeper in holes that I avoided, where the gators lay. There were hundreds of orchids, clamshells and night-scented, rigid and dingy (poor fellow isn't dingy at all, just subtle), a reestablished cigar orchid doing well in its new home. We even spotted ghost orchid roots once, but could never find them again. Orchids clung to lichen-mottled trunks of pop ash and pond apple, oak and cypress, and hid in emerald and tawny nooks. Clamshell orchids in their Elizabethan cloaks and night-scented orchids, shamelessly exposing their reproductive parts, were in magnificent bloom. Dozens of butterfly orchids were ready to explode with blossoms. This place is redolent of the jungle—but how did these improbable tropical species get here? Why are they so happy in this slough, where spirits dwell?

If you look at a map, it's obvious. The tip of Florida is the northern outpost of the Caribbean range, and the seeds of most native orchids probably arrived on the wind. The habitat was favorable—warm and damp, with distinct wet and dry seasons—and the associated fungi either came along or evolved to fill the niche.

Some pollinators probably came the same way—they blew in with the breeze—but others may never have arrived at all.

Form and Function

Scientists characterize orchids as monocots, which means they are seed plants with a single seed leaf, parallel leaf veins, and flower parts in threes. Though extremely diverse, the 25,000 or more species in the Orchidaceae, and five times as many hybrids, share certain flower structures. All have three sepals, or outer whorls, and three petals, or inner whorls. One of the petals is showier, and this lip, or labellum, is usually the lowest segment and forms a landing pad for pollinating insects.

Orchids are undeniably erotic. They are both male and female in form. Projecting from the center is the fleshy, club-shaped column, a fusion of the male (staminate) and female (pistillate) reproductive organs. At the apex of the column is the anther with pollen grains clustered in masses of two to eight, called *pollinia*. Nearby is the female portion, the stigma, ready to receive the pollinia and absorb it into the ovary which, after fertilization, expands into a seed capsule. Ah, but it is rarely that simple . . .

85

Strange Strategies

Orchids cannot walk, run, crawl, swim, fly, or leap to meet their lovers. Instead, to get male pollen and female ovaries together, orchids have learned—evolved—to trick mobile creatures into serving them, sometimes by offering nectar rewards. These orchids specialize in "bait and switch." Charles Darwin devoted years to the study of orchids' tricks and the partners with which they evolved for successful reproduction, but only in the last few years has the scientific community shown a respect for the deceptive powers of orchids. Since Darwin's time, 3,000 orchids have been identified worldwide that use sexual deception to attract pollinators. Appearance and odor play the major roles.

Orchids are gymnasts, too. Their flowers perform as they emerge, full of tricks, or so it seems. The buds of many species begin to open upside down—lip up—but, while they are enlarging, they turn 180 degrees and fully open with the lip lolling languidly at the bottom, ready for seduction. Science calls the process *resupination*—whatever it is the poor pollinators have no chance! The orchids are irresistible.

Sex and the Stupid Bee

Maybe males really are all alike—sex on their minds all the time. Certainly, if euglossine bees have a mind, sex is all that is on it. They are driven to beelike *Cyrtopodium* by the flowers' good looks and, in most cases, by their sexy perfume. So specific are the two—blossom and bee—that euglossine bees are also known as orchid bees, and *Cyrtopodium punctatum*, found in southern Florida, is also known as the bee-swarm orchid.

The bee is driven to a specific orchid flower in lust and tries to mate with her. He probes and pushes and never quite gets it done. Finally, frustrated, he withdraws to find another lover, but in so doing, picks up the pollinia on his legs or head. Off he flies to another potential mate,

probing and pushing again, and deposits the pollinia into the anther of the new plant. The satisfied orchid subtly changes its perfume from "come and get it" to "not tonight, not this year." Meanwhile, the bee persists in this pseudo-copulation until he is exhausted, or, if his species is lucky, he finds a female of his own kind. He may be too exhausted to try again, leaving female euglossine bees without progeny.

Some say that Florida's *Cyrtopodium* did not bring its euglossine bee with it from the Caribbean. However, *Cyrtopodium* has managed to get pollinated and reproduce in Florida for centuries, so bees must be doing it. I once saw several of what I thought were euglossine bees way up, harassing the outer blossoms of a huge *Cyrtopodium* wrapped around a cabbage palm. Maybe they were plain old bees, curious. We do know that unless they are hand pollinated, the bee-swarm orchid usually blooms in vain.

A bee compulsively probes and jostles to get inside a blossom.

Larry Richardson hand-pollinates a cigar or bee-swarm orchid (Cyrtopodium punctatum).

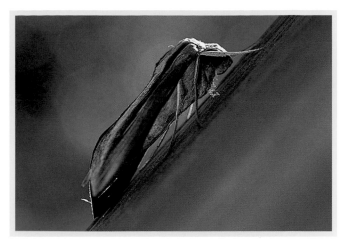

The banded sphinx moth (Eumorpha fasciata), luxuriates in an improbably long proboscis which fits perfectly into the elongated nectar spur of the ghost orchid (Dendrophylax lindenii) to effect pollination.

Where Are They Now?

"Where have the pollinators gone? It's a mystery," said Larry rhetorically, as he used tiny forceps to remove the pollinia from the anthers of several flowers on a bee-swarm orchid. He deposited them in the stigma of flowers on another plant, labeling each to keep track of the genetic mix and to monitor the success of his work. He repeated this process among several widely dispersed plants.

Where are the bees now? Have they been poisoned by pesticides used on the agricultural fields pressing in from the north and from the west? Has their habitat been critically altered by the shortened hydroperiod? Are the plants, in fact, too scattered for natural pollination? We need to know much more.

There are huge egos and conflicting theories about orchid pollination. Some, such as the night-scented orchid, are thought to have evolved to self-pollinate here in Florida. If this is true, I wonder why its sweet scent is released only after the sun sets, and why the flower is white —usually a tip-off for night pollination. There must be a moth to pollinate it, unless the moth was left "back home" in the Caribbean.

They say the ghost orchid is pollinated by the gigantic sphinx moth, which is equipped with a proboscis long enough to slip down the orchid's three-to-six-inch nectar spur. The ghost rewards the moth with the nectar, so that, while indulging, the moth will catch the pollinia on its head and transfer it to another bloom. Win Turner says a friend who was also born and raised in the south Florida swamps told him he had witnessed ghost orchid pollination. I would like to see for myself, but so far I have failed. I have set up elaborate camera traps before the open mouths of ghost orchids, but I have never captured a pollinator on film. I have seen innumerable mosquitoes perched on ghost orchid blooms, and I have photographed resting Sphinx moths, but never have I caught them in the act. I am still trying.

I once hung in a tree for half an hour photographing a mosquito hawk poking all over a tiny frosted flower orchid (Pleurothallis gelida). John Beckner dismissed my proud photo presentation. "That's not the pollinator," he said. I was crestfallen. What was that insect doing there all that time? What, then, is the pollinator? "We don't know," announced John, "but that's not it."

Farming Fungus

Once pollinated, the ovary forms a pod with a million or more powder-light seeds that disperse with the wind and, perchance, meet up with the

Mycorrhizal fungus for Cyrtopodium punctatum *is being grown in a petri dish. One cubic centimeter will be placed in other petri dishes and readied for introduction of orchid seeds. Scott Stewart and Larry Richardson are working to understand the science of wild orchid propagation.*

right fungi (or, as Mike Owen would say, "fun guy"), attach to the perfect tree, and live happily ever after. But again, it's not that simple.

Most North American orchids obtain their sustenance by using chlorophyll for photosynthesis, even those plants with no leaves. However, all orchid species also trick fungi into feeding them. It probably started early. Orchids had an abundance of seeds that were attacked by fungus because that is what fungus does—breaks down organic matter. As the fungus broke through the outer shell, it released the seed to germinate. Eventually the orchid seeds could not germinate without the aid of the fungus, and the fungus, remaining in the roots as the orchid grew, became more and more dependent and could not live without the orchid, which feeds it.

Orchid seeds themselves have no source of energy. They are just lumps of embryonic cells with a one-cell-thick, nearly transparent seed coat. They depend on the attack of the fungi for germination. The orchid embryo quickly turns the tables on the predatory fungus and begins to digest it, eating the carbohydrates and amino acids, and absorbing the water from the fungus. This act of fungus eating, termed *mycotrophy*,

is what powers the minute orchid embryo through germination and early development. Once a green leaf appears, photosynthesis can begin, but the fungus will remain trapped in the orchid roots until the orchid decides it needs an extra boost of mycotrophic energy for a growth spurt or for reproduction.

These orchid-mycorrhizal fungal associations have been known for some years, but not enough intensive research has been done to fully understand the relationship. Scott Stewart is changing that. Working with colleagues at the University of Florida and at the Smithsonian Institution, he is trying to uncover the unique mycorrhizal fungus for each of Florida's native orchids. If successful, he anticipates quick reestablishment in the wild.

Anyone can flask orchids, sprinkling seeds into an agar medium and growing seedlings, but when the plants are taken out of the sterile environment and put into a greenhouse they suffer a 95 percent mortality rate. Scott expects to reverse that number, saving 95 percent. If the orchid germinates with its mycorrhizal fungus, when it is put into the wild it already has a source of nutrition to give it a boost while taking root and establishing its photosynthetic activity.

Though he is a scientist, Scott sees miracles. "If the seed and the fungus match, a miracle occurs almost overnight. The fungus will infect the seed and spur a rapid transformation from seed to plant—it is really quite amazing to see firsthand."

We still do not know why orchids proliferate in certain areas and are absent from others that look equally enticing. We don't understand why some species exult on low cypress trunks in housing developments and near strip malls, stores, and large service organizations. Is it the existence of the mycorrhizal fungi that gives them a foothold? Is it the pollinators, prevalent in one area and not another, that dictate where orchids thrive and where they don't? The strands hold the answers to many mysteries, but they reveal their secrets slowly.

Larger-than-life flowers command attention and reduce the distance between nature and the human soul. They draw people into a parallel universe of complex, rooted, living things active in their own procreation. Clamshell orchid (Prosthechea cochleata var. triandra*).*

Born in the deepest swamps, the diminutive, erect, perky little pom-poms, of the so-called dingy orchids (Epidendrum amphistomum) *are private, but their sensual beauty, inviting the viewer in, is unnervingly erotic.*

In less than a year, this and other rescued night-scented orchids (Epidendrum nocturnum) had established roots and clung to limbs, allowing biologist Larry Richardson to remove the ties.

Though it is endangered, the night-scented orchid (Epidendrum nocturnum) is found throughout the swamps and hammocks of south Florida, the West Indies, Mexico, Central America, and northern South America. White suggests that it attracts night moths, as does its pungent fragrance, released after the sun's radiance withdraws.

Cigar or bee-swarm orchid (Cyrtopodium punctatum)
"I am dizzy with pleasure at the beauty, natural beauty, perfection in form, of this gift. But who sees? Few. Very few. She dances alone for a lover who will never visit her again. Without us, humans, our magniicent orchid would flower and wither and flower and wither without ever producing another of its own kind."

—Connie Bransilver and Larry W. Richardson
Florida's Unsung Wilderness: The Swamps

Rare hermits, Epidendrum strobiliferum *retreat to the deepest reaches of the sloughs, establishing dense but inconspicuous tangles of stems, leaves, and modest pale yellow flowers high up in the trees.*

Red and green, complementary colors, make a powerful impact in miniature proportions.

Fakahatchee ladies'-tresses (Sacoila lanceolata *var.* paludicola) *require intimate relationships.*

Plants of Polystachya concreta *are not common, but they are widespread. Epiphytic, resting on a variety of trees, their flowers are like a head of unruly, curly, blond hair.*

Ghost Orchid (Dendrophylax lindenii). *A greenhouse ghost orchid—if there were a captive ghost orchid—would be to an untamed wild ghost as a house cat is to a panther. Like the panthers that share their habitat, ghost orchids have never become domesticated. Ghosts, especially, are still savage and independent, living and seducing their life-giving pollinator in a universe we can only dimly perceive.*

*Delicate ionopsis (*Ionopsis utricularioides) *is often found hovering above the water.*

Still confused, Liparis elata's *pods poke straight upward on the stalk, not down like other pods.*

Epidendrum rigidum, *one of the most common orchids in the swamps, blends well with its surroundings.*

A naturalized South American, yellow cowhorn (Cyrtopodium polyphyllum) is *terrestrial, while its cousins are epiphytic. Like the cigar orchid (Cyrtopodium punctatum),* it is a flood of golden sunlight. This species is found only in Dade County.

It is said that less than 1 percent of the world's flora has been thoroughly studied by science, and the dwarf butterfly orchid (Prosthechea pygmaea) is one reason. Not only is this miniature easily overlooked, but most of the flowers don't fully open. We know so little, understand so little.

While some pine-pinks (Bletia purpurea) open to the sunlight and pollination, other flowers on the same plant never fully open. Like many night-scented orchids they can be self-pollinating. The anther bends and deposits its pollinia directly into the stigma, effecting pollination.

Every breeze makes the clamshell orchid (Prosthechea cochleata var. triandra) *seem to dance for unseen lovers. In fact, that is not far from the truth. Devoid of the profound fragrance found in some of its wild orchid neighbors, it must attract its insect pollinators with showy colors alone.*

CHAPTER 5
Conservation and Restoration

"We all need a place on Earth to hide the wild parts of ourselves. With land preserved, each generation can search again . . . and discover . . . a greater sense of who we are."

—*Jesse Taylor-Ide*
Shadows in the Sun *by Wade Davis*

Loving Them to Death

"It's chaos out here!" shouts Mike Owen, as he sprints through sword fern, lianas set to ensnare him, rotting logs, cypress knees, and water—everything jumbled wildly together, plants competing for light in this jungle. He is the picture of pure glee, winding up to manic proportions, singing and hooting and shouting the swamp's praises. Mike is trying to take the pulse of this magical place, hidden under a sacred veil of mist and heavy air. He sees the wild orchids as its savior, endowed with the power to touch peoples' hearts and minds and promote changes in attitudes and policies.

೧೦

The urge to conserve is prompted much of the time by a spiritual connection with nature, but greater understanding of the science of conservation can only strengthen our appreciation of and commitment to it. To save orchids and their natural habitat, more people have to see them in the wild, to contemplate not only their beauty and their complex life histories, but also their interrelationships with swamp, forest, and prairie. Most people in this country claim to be "conservative" but many somehow fail to make the connection between conserving a culture and conserving what has preceded all human culture—the places of natural beauty where spirits dwell.

Wetland Water Flow

People move to southwest Florida for the weather, the beaches, the "good life in paradise." Unbeknownst to most residents of the Sunshine State, all of these things are tied to the unique and still somewhat pristine environment of Florida's middle, its heart. Few understand this dependence or appreciate the life-giving system of inland swamps and marshes. So the water is dammed and diverted, its flow blocked and siphoned into huge canals that spill out into the 10,000 Islands and Florida Bay, changing the salinity and making fish nurseries sterile and devoid of sea life.

This is how the system *should* work: Water from Lake Okeechobee, the third largest fresh-water lake in the U.S., runs south and east into the Everglades' saw grass prairies, Lilliputian cypress, and hardwood hammocks to the mangroves and Florida Bay. It also runs west and south, creating, along with natural precipitation, the Big Cypress Basin of which the various wooded strands and sloughs—Fakahatchee Strand, Camp Keais Strand, Corkscrew Swamp, Okaloacoochee Slough—are all a part. That water, too, runs into Florida Bay through the 10,000 Islands. As it commutes to the bay at a rate of a mile a day, much of the water percolates through to the underground aquifers—which we tap for our drinking water—cleaning itself in the

Spotting Frosted flower (Pleurothallis gelida) *plants in bloom among the cacophony of greens and reflected lights in the swamp is nearly impossible.*

process. The wetlands, if left intact, would be a perfect filtration and flood-control system.

Unfortunately, we have diverted the water into dikes and canals that flow too fast to recharge the freshwater aquifers. We've drained the swamps, introduced invasive species, cut big trees and mangroves, and cleared native plants for farms and houses in the midst of tropical paradise. But we cannot fool with Mother Nature. When the seasonal rains come—an average year dumps more inches of rain on south Florida than I am tall—she reclaims her stomping grounds. Nature's abused past means a perilous future for native orchids.

The solution is to restore historic water flow to these unique wetlands. Native orchids are known as an umbrella species or charismatic microflora. This means that saving habitat for the orchids—beautiful flowers with which ordinary people can identify—also saves the myriad and perhaps less charismatic flora and fauna that share their habitat. To appreciate and then respect the natural ecosystem, people must connect with these wild lands on many levels—intellectual, emotional, aesthetic, practical and spiritual—and also to the places already crippled by our greed. Orchids lead us. They help us find humanity's relationship with the natural world and a harmony of our sensibilities. We have a responsibility to our progeny—those beautiful grandchildren whose pictures are whipped out of wallets all over south Florida—and to the land itself to understand and conserve this rugged, spiritually enriching natural treasure. We must listen to our vital core, our hearts and our heartland, and do whatever it takes to maintain this paradise.

Research and Restoration

We call these places wild—this checkerboard of federal, state, county, and private lands—not because they are untouched, virgin, but because they have not been brought completely to heel under our Western rules of polished lawns and geometric shrubs. For the lower 48 states, the Florida swamps are about as wild as it gets.

It may be too late for many of Florida's beleaguered orchids. Urban sprawl is gnawing at the edges of the swamp and pushing through lax regulations to eat more ravenously at wild-lands. While orchids leave no carcasses for vultures, and we do not gasp at the carnage as we whip past in our SUVs, they are no less victims of our expansion. Not only do plants die, they are never born because their life-giving tree limbs have disappeared. Their cries for help are silent, like Munch's "The Scream"—noiseless to the ears, but deafening in the heart—harbingers of what may be the future for all living things. Orchids, panthers, and other voiceless creatures can only retreat and weep.

Natural water flow is the evident and only solution. Yet eager fingers of human habitation push farther and farther into the paths of that natural flow, emboldened by ineffectual local regulations that crumble under the development juggernaut loosed by a lax national commitment to conservation. While wild orchids incidentally benefit from the focus on intact biosystems, few ecosystems have been saved solely because of the existence of charismatic microflora. The problems seem overwhelming; however, there may be hope.

We are on the cusp of promised Comprehensive Everglades Restoration Plan, although the project itself is under siege as big-money interests in agriculture, sugar, and development take aim at its $8-billion budget. As passed by Congress and signed by the President, it will be the largest environmental effort in the world to date. Though an imperfect plan, riddled with contradictions and competing interests, it is the best hope we have of restoring what we can of the original hydrology of this unique wilderness.

We are seeing a collapse in orchid populations, not just a scarcity. Many species are already technically extinct, in that existing plants are only living out their life spans, not reproducing. This urgency calls not only for conservation, which may be enough in some areas, but for more drastic measures: scientific restoration of historic orchid populations.

Fortunately, zoological societies around the world have shifted their focus from collecting specimens to supporting wild populations in their natural habitats. Botanical societies largely have not. Conservation of native orchid species in their natural habitat is still the poor stepchild of orchid societies which, to this day, dote on vast, mostly private collections of rare orchid species and their hybrids. Modern zoologists and botanists were left with the same legacy of empire —large collections of rare and endangered species, many quite

desiccated, some living, if not thriving, in captivity, and a few species surviving only because they were captured before they became extinct in the wild. The American Orchid Society's (AOS) Conservation Committee has supported awards, grants, and orchid rescue attempts along with informational and educational materials to aid conservation, but, again, their definition of conservation seems to focus on preserving specimens in collections rather than saving key orchid habitats.

Captive propagation of endangered orchids has its place. Probably the most effective conservation effort of AOS has been to work with the U.S. Fish and Wildlife Service to better understand and remedy the deficiencies of CITES (Convention on International Trade in Endangered Species) as it pertains to the importation of endangered orchid species. In general, CITES prevents the propagation of threatened orchid species outside the country of their origin. The reality is that priorities are different in many host countries, where the immediate need to feed hungry mouths ranks way above propagating endangered non-edible plants. Burgeoning populations needing immediate governmental

An ABC News team covered the important story of hand pollination of cigar orchids in an effort not only to save these remarkable native orchids but also to call attention to the fragility of their habitat, Florida's forgotten western Everglades.

services are expanding into wildlands, and orchid appreciation is often a rich country's indulgence. However, some poorer nations are beginning to see the potential in orchid ecotourism and bargaining for investments in the preservation of forests and wetlands.

The AOS offers to assist other groups, such as The Nature Conservancy, National Audubon Society, and World Wildlife Fund, to manage significant orchid habitats. It is a step in the right direction. It seems time for the old habits and perverse territorial imperatives of the past to give way to more robust and sustainable management methods—*in situ* conservation and, where needed, restoration.

While the science proceeds on one hand, a dedicated nucleus of orchid lovers has rallied supporters to bring the plight of Florida's native orchids to public attention. Lee Hoffman, president of the Native Orchid Restoration Project, along with other enthusiasts, has established a master plan for orchid restoration. With imagination and plans that would require a small army to accomplish, Lee plunges forward with his dream—orchid seeds propagated and seedlings and rescued orchids reestablished in the swamps from which they came. Native Floridians are being interviewed to determine what plants existed, where, in what numbers, and in what habitat. Research on hydrology, humidity, and temperatures is coming in from Big Cypress. By partnering with public, scientific, and conservation entities as well as private organizations and individuals, NORP hopes to create a common collective knowledge on orchid conservation that will be available to all—tall order.

Nascent conservation and restoration projects are underway all over the world. Genetic studies in Great Britain are seeking the nearest relatives of a nearly extinct species of terrestrial orchid, ecotours and propagation may help to save a vanishing culture of orchids in Malaysian Borneo, and one dedicated man in Guatemala has rescued thousands of orchids. Caught between conservation and development, Otto Mittelstaedt rescued orchids from felled trees, grew them, described them, and found 42 new species along the way. In fact, it appears that the greatest success in orchid conservation occurs when one or several dedicated human beings, enthusing others, have changed their piece of the world, proving, again, Margaret Mead's words: "Never doubt that a small group of committed citizens can change the world. Indeed, it is the only thing that ever has." These individuals, seemingly rare in the orchid world, invert the proverbial orchid collectors' obsession and greed, and instead, want to share, to give back. They are the antithesis of many European explorers who ripped plants and animals from their native lands to bolster their egos and the competitive standing of the Motherland.

One thing is certain, however. The depletion of orchids is an indicator of the changes set in motion by our past mistakes. Reinvigoration of orchid populations can similarly be an indicator of our learning. Dr. Jane Goodall, DBE, who is always a sane voice for conservation, teaches that we cannot ignore the consequences of our actions. Many of her most gripping stories about the chimpanzees of Gombe, which she has studied for 40 years, involve the chimps' drive toward domination over and control of their peers. It is possible that our hunger for power pre-dates even those three million years that humans were separate from other apes. The lust for control over the wildest and rarest must run deep. But, Jane's mantra is hopeful: Each individual makes a difference. We, unlike our fellow apes, can think through our actions and make positive impacts on the world around us.

The notorious ghost orchid (Dendrophylax lindenii) stars in courtroom dramas, books, and the movie Adaptation. *This plant's current bloom, springing from its distinctive star root pattern, nods above last year's still-ripening seed pod.*

Diminutive flowers of the dingy-flowered star orchid (Epidendrum amphistomum) *perch on erect stems, which themselves rest, as epiphytes, on a large variety of trees. These pale flowers, overlooked by those seeking intense color, often go unappreciated.*

A perfectly colored spider in miniature spins its trap for even tinier prey, transmuting its catch into sustenance and survival and linking space and time. Cigar orchid (Cyrtopodium punctatum).

Next Page: *Fierce lightning storms during the rainy season sweep across flat south Florida.*

The first shoots of Encyclia tampensis *are one-half to three-quarters of a centimeter long. This shoot was sent to the Smithsonian Environmental Research Center outside Washington, D.C., for study of the plant's mycorrhizal fungi.*

*In a world of a thousand shades of green, fading up to blues and down to browns, miniature mirrors of white light catch my eye. The white butterfly orchid (*Encyclia tampensis *forma* albolabia*) is rarely seen. A chance to caress this haunting variant delights the eye and heart.*

Gentian noddingcaps (Triphora gentianoides) *sneak into mulch and erupt in gardens in south Florida. It is almost impossible to find the flowers fully open. As I bent down to photograph the only one open, a mere three inches off the ground, its stem broke. Never mind. I placed it in the center of a nearby bromeliad, making a colorful studio to frame* Triphora.

The ribbon orchid (Campylocentrum pachyrrhizum) *is one of only three leafless epiphytic orchids in the United States. The others, both found in south Florida, are the ghost orchid* (Dendrophylax lindenii) *and the jingle bell orchid* (Harrisella porrecta). *Its roots contain more visible chlorophyll than either of the others.*

Big, boisterous, and bright yellow Oncidium floridanum *is terrestrial with large pseudobulbs.*

It is said that Harrisella porrecta *produces a jelly that kills the branch on which it rests in order to make more light before the branch falls.*

Once thought to be endemic to southern Florida, the delicate Florida star orchid (Epidendrum floridense) sparsely inhabits south Florida swamps and nearby Cuba as well. This tiny plant was found, uncharacteristically, on a royal palm.

You have to be twisted to search for the twisted or cone-bearing epidendrum (Epidendrum strobiliferum). That's what I thought, anyway. For several hours, I had dragged tripod and camera gear through deep water infested with bladderwort's yards and yards of foot- and tripod-grabbing stems. The orchid itself was high above, so the tripod was useless as it sank in the mud below, as did I. The Outward Bounders with me saved the day. One steadied me, while others kept cords and photo equipment out of the water.

Afterword

"In the woods, we return to reason and faith. There I feel nothing can befall me in life—no disgrace, no calamity . . . which nature cannot repair."

—*Ralph Waldo Emerson*

Those of us who care about wilderness do not have the luxury of ignoring the growing pressure of booming population on the natural world. There is a tidal wave of humans moving into Florida, and the population in southern Florida has increased more than 85 percent since 1970. Every assault on our wild places and therefore on wild orchids—markers of our health and harbingers of our future—is an assault on our own wellness. Our sheer numbers precipitate habitat loss due to logging and other extractive industries, alteration of the natural water flow and the hydroperiod, poaching, pesticides, fragmentation of orchid populations, and, worst of all, urban sprawl.

In a society that seems bent on doing away with the natural world, taming and homogenizing every living thing, it is refreshing to enter the swamp. Its allure is subtle. Shunned and misunderstood, it still creates fire in my heart. To me, the "forgotten" western Everglades is a place of solitude and compares well with other great tropical forests I have experienced in Borneo, Sulawesi, West Africa, Madagascar, the Amazon, and Daintree in Australia. Yet this jungle is right here, shyly offering her diversity and beauty to those with a sense of adventure and respect. Her special treats are the orchids, demure or brazen but always mysterious, which reveal their secrets slowly. There is so much more to learn . . .

A bird wades on the edge of a slough.

Huge, blooming cigar or bee-swarm orchid (Cyrtopodium punctatum) *plant on cabbage palm tree.*

The Naming Conundrum

"We should have a common-name congress every year to think of new common names for all the orchids," proposed Mike Owen, not entirely in jest. "Jingle bell, elephant ear, roller coaster, frog . . . much more fun!" Mike, like the rest of us nontaxonomists, having finally committed to memory the Latin names used in Carlyle Luer's book, *The Native Orchids of Florida*, is frustrated by having to learn more recent nomenclature used by Paul Martin Brown in his book *Wild Orchids of Florida*. The International Association of Plant Taxonomists meets every five years to set rules under which names are chosen. Academics look down their noses at those of us struggling to make sense of these names, but they are right in saying that common names are useless, as they differ so widely for the same species.

Luer's seminal work, published in 1973, identified and displayed photographs of several species no longer found. For years this book was the bible of Florida orchid identification, and copies of the volume, if they can be found, now sell for well over $200. Though southwestern Florida still offers the greatest diversity of native orchids anywhere in the United States, too many of the 68 species documented by Luer and still listed by the Florida Department of Agriculture have been missing for many years. Herein lies the tragedy.

Paul Martin Brown's book incorporates the latest understanding of the classification of orchid species. He has added four species and one variety to the existing list for Florida. He discovered four species new to science and described, redescribed, or reclassified a number of new varieties or forms. The regal clamshell orchid, for example, used to be *Encyclia cochleata*, but now it is *Prosthechea cochleata*. Oh, well. Just learn it and all the others, too. I use the Brown nomenclature throughout this book.

The pods of Harrisella porrecta *give it one of its common names, jingle bell orchid.*

Orchids of Southern Florida

Compiled by Paul Martin Brown

Covering Lee, Hendry, Collier, Monroe, Broward, and Miami-Dade Counties

† Known in the United States only in the southern Florida counties above

* Presumed extirpated

Basiphyllaea corallicola (Small) Ames †
Carter's orchid

Beloglottis costaricensis (Reichenbach *f.*) Schlechter †
Costa Rican ladies'-tresses

Bletia purpurea (Lambert) de Candolle
pine-pink

Brassia caudata (Linnaeus) Lindley †*
spider orchid

Bulbophyllum pachyrachis (A. Richard) Grisebach †*
rat-tailed orchid

Calopogon barbatus (Walter) Ames
bearded grass-pink

Calopogon multiflorus Lindley
many-flowered grass-pink

Calopogon pallidus Chapman
pale grass-pink

Calopogon tuberosus (Linnaeus) Britton, Sterns
  & Poggenberg var. *tuberosus*
common grass-pink

Calopogon tuberosus (Linnaeus) Britton, Sterns
  & Poggenberg var. *simpsonii* (Small) Magrath †
Simpson's grass-pink

Campylocentrum pachyrrhizum (Reichenbach *f.*)
  Rolfe †
crooked-spur orchid; ribbon orchid

Corallorhiza wisteriana Conrad
Wister's coralroot

Cranichis muscosa Swartz †*
moss-loving cranichis

Cyclopogon cranichoides (Grisebach) Schlechter
speckled ladies'-tresses

Cyclopogon elatus (Swartz) Schlechter *
tall neottia

Cyrtopodium punctatum (Linnaeus) Lindley †
cowhorn orchid; cigar orchid

Dendrophylax lindenii (Lindley) Bentham *ex* Rolfe †
ghost orchid; frog orchid

Eltroplectris calcarata (Swartz) Garay & Sweet
spurred neottia

Encyclia tampensis (Lindley) Small
Florida butterfly orchid

Epidendrum acunae Dressler †*
Acuña's star orchid

Epidendrum amphistomum A. Richard
dingy-flowered epidendrum

Epidendrum floridense Hágsater
Florida star orchid

Epidendrum nocturnum Jacquin
night-fragrant epidendrum

Epidendrum rigidum Jacquin
rigid epidendrum

Epidendrum strobiliferum Swartz †
cone-bearing epidendrum

Eulophia alta (Linnaeus) Fawcett & Rendle
wild coco

Galeandra bicarinata G. A. Romero
  & P. M. Brown †
two-keeled galeandra

Govenia floridana P. M. Brown †*
Florida govenia

Habenaria distans Grisebach
false water-spider orchis

Habenaria odontopetala Reichenbach *f.*
toothed rein orchis

Habenaria quinqueseta (Michaux) Eaton
Michaux's orchis

Habenaria repens Nuttall
water-spider orchis

Harrisella porrecta (Reichenbach *f.*) Fawcett & Rendle
leafless harrisella

Hexalectris spicata (Walter) Barnhardt var. *spicata*
crested coralroot

Ionopsis utricularioides (Swartz) Lindley
delicate ionopsis

Lepanthopsis melanantha (Reichenbach *f.*) Ames †★
crimson lepanthopsis

Liparis elata Lindley
tall twayblade

Macradenia lutescens R. Brown †★
Trinidad macradenia

Malaxis spicata Swartz
Florida adder's-mouth

Maxillaria crassifolia (Lindley) Reichenbach *f.* †
false butterfly orchid

Maxillaria parviflora (Poeppig & Endlicher) Garay †★
small-flowered maxillaria

Mesadenus lucayanus (Britton) Schlechter
copper ladies'-tresses

Oncidium floridanum Ames †
Florida oncidium

Pelexia adnata (Swartz) Sprengel †★
glandular ladies'-tresses

Platanthera nivea (Nuttall) Luer
snowy orchis

Platythelys sagreana (A. Richard) Garay
Cuban ground orchid

Pleurothallis gelida Lindley †
frosted pleurothallis

Polystachya concreta (Jacquin) Garay & Sweet
yellow helmet orchid

Ponthieva brittoniae Ames †★
Mrs. Britton's shadow-witch

Ponthieva racemosa (Walter) Mohr
shadow-witch

Prescottia oligantha (Swartz) Lindley †
small-flowered prescottia

Prosthechea boothiana (Lindley) W. E. Higgins
  var. *erythronioides* (Small) W. E. Higgins
Florida dollar orchid

Prosthechea cochleata (Linnaeus) W. E.
  Higgins var. *triandra* (Ames) W. E. Higgins
Florida clamshell orchid

Prosthechea pygmaea (Hooker) W. E. Higgins †
dwarf butterfly orchid

Pteroglossaspis ecristata (Fernald) Rolfe
crestless plume orchid

Sacoila lanceolata (Aublet) Garay var. *lanceolata*
leafless beaked orchid

Sacoila lanceolata (Aublet) Garay var. *paludicola*
  (Luer) Sauleda, Wunderlin & Hansen
Fakahatchee beaked orchid

Spiranthes amesiana Schlechter *emend.* P. M.
  Brown †★
Ames' ladies'-tresses

Spiranthes brevilabris Lindley
short-lipped ladies'-tresses

Spiranthes eatonii Ames *ex* P. M. Brown
Eaton's ladies'-tresses

Spiranthes laciniata (Small) Ames
lace-lipped ladies'-tresses

Spiranthes longilabris Lindley
long-lipped ladies'-tresses

Spiranthes odorata (Nuttall) Lindley
fragrant ladies'-tresses

Spiranthes praecox (Walter) S. Watson
giant ladies'-tresses

Spiranthes torta (Thunberg) Garay & Sweet †
southern ladies'-tresses

Spiranthes vernalis Engelmann & Gray
grass-leaved ladies'-tresses

Trichocentrum carthagenense (Jacquin) M. W.
  Chase & N. H. Williams †★
spread-eagle orchid

Trichocentrum undulatum (Swartz) Ackerman
  & M. W. Chase †
spotted mule-eared orchid

Triphora craigheadii C. A. Luer
Craighead's noddingcaps

Triphora gentianoides (Swartz) Ames &
  Schlechter
Least-flowered triphora; gentian noddingcaps

Tropidia polystachya (Swartz) Ames †
many-flowered tropidia

Vanilla barbellata Reichenbach *f.* †
worm-vine; leafless vanilla

Vanilla dilloniana Correll †★
Dillon's vanilla

Vanilla mexicana Miller
scentless vanilla

Vanilla phaeantha Reichenbach *f.* †
oblong-leaved vanilla

Naturalized, Introduced, Escaped & Waifs

Bletia florida (Salisbury) R. Brown †
slender pine-pink

Bletia patula Hooker †
broad-lipped pine-pink

Cyrtopodium polyphyllum (Vell) Pabst *ex* F. Barrios †
yellow cowhorn orchid

Epidendrum cf. *radicans* Paven *ex* Lindley †
flame star-orchid

Laelia rubescens Lindley †
pale laelia

Oeceoclades maculata (Lindley) Lindley
African spotted orchid

Vanilla planifolia Jackson *ex* Andrews †
commercial vanilla

Vanilla pompona Scheide †
showy vanilla

Zeuxine strateumatica (Linnaeus) Schlechter
lawn orchid

Pine-pinks (Bletia purpurea) *are sometimes not pink at all, but white with golden interiors* (Bletia purpurea *forma* alba).

Bibliography

Ames, Oakes. 1948. *Orchids in Retrospect: A Collection of Essays on the Orchidaceae. A Faithful Reproduction of the 1948 Classic with an Appreciative Foreword by Richard Evans Schultes*. Stanfordville, NY: Botanical Museum of Harvard University. 1979. Earl M. Coleman.

Balog, James. 1999. *Animal*. New York: Graphis Press.

Bayman, Paul, Ph.D., et al. July 2003. "Hidden Transactions: The Curious Relationships between Orchids and Fungi." *Orchids*. Delray Beach, FL: American Orchid Society.

Berliocchi, Luigi. 1996. *The Orchid in Lore and Legend*. Translated by Lenore Rosenberg and Anita Weston, edited by Mark Griffiths. 2000. Portland, OR: Timber Press

Bransilver, Connie, and Larry W. Richardson. 2000. *Florida's Unsung Wilderness: The Swamps*. Englewood, CO: Westcliffe Publishers.

Brown, Paul Martin, with drawings by Stan Folsom. 2002. *Wild Orchids of Florida with References to the Atlantic and Gulf Coastal Plains*. Gainesville: University Press of Florida.

Brown, Paul Martin, Drawings by Stan Folsom. 2003. *The Wild Orchids of North America, North of Mexico*. Gainesville, FL: University Press of Florida.

Davis, Wade. 1998. *Shadows in the Sun*. Washington, DC: Island Press.

Dimock, A. W. 1915. *Florida Enchantments*. Peekamose, NY: Outing Publishing Company.

Duever, Michael J., et al. 1979. *The Big Cypress National Preserve*. New York: National Audubon Society.

The Everglades Rises Again. Audubon July–August 2001 special issue. New York: National Audubon Society.

Fitch, Charles Marden. May 2003. "Fragrant Encyclias." *Orchids*. Delray Beach, FL: American Orchid Society.

Florida Fish and Wildlife Conservation Commission. *Florida's Endangered Species, Threatened Species and Species of Special Concern*. 1 August 1997. www.floridaconservation.org/pubs/endanger.

Flowers, Mark, producer. 2000. *Obsession with Orchids*. Thirteen/WNET New York, NY, and BBC-TV, London.

George, Jean Craighead. 1995. *Everglades*. New York: Harper Collins.

Hagsater, Eric, and Vinciane Dumont, eds. Compiled by Alex M. Pridgeon. 1996. *Orchids: Status Survey and Conservation Action Plan*. IUCN/SSC Orchid Specialist Group, IUCN, Gland, Switzerland and Cambridge, UK.

Hammer, Roger L. 2002. *Everglades Wildflowers: A Falcon Guide. A Field Guide to Wildflowers of the Historic Everglades, Including Big Cypress, Corkscrew, and Fakahatchee Swamps*. Guilford, CT: The Globe Pequot Press.

Hansen, Eric. 2000. *Orchid Fever: A Horticultural Tale of Love, Lust, and Lunacy*. New York: Pantheon Books.

Kloor, Keith. July–August 2001. "Forgotten Islands" *Audubon*. New York: National Audubon Society.

Koopowitz, Harold. 2001. *Orchids and Their Conservation*. Portland, OR: Timber Press

Luer, Carlyle A. 1972. *The Native Orchids of Florida*. New York: New York Botanical Garden.

Orlean, Susan. 1998. *The Orchid Thief*. New York: Random House.

Ripple, Jeff. 1996. *Southwest Florida's Wetland Wilderness*. Gainesville: University Press of Florida.

Saint-Exupery, Antoine de. 1943. *The Little Prince*. Translated from the French by Katherine Woods. London: Harcourt Brace & Co.

Schiestl, Florian P., Ph.D., and Rod Peakall, Ph.D. October 2002. "Floral Odor and Insect Olfaction." *Orchids*. Delray Beach, FL: American Orchid Society.

Sheehan, Thomas J. 2001. *Ultimate Orchid*. New York: DK Publishing

Simmons, Glen, and Laura Ogden. 1998. *Gladesmen: Gator Hunters, Moonshiners and Skiffers*. Gainesville: University Press of Florida.

Stewart, Scott. October 2002. "Saving a Native Orchid." *Orchid*. Delray Beach, FL: American Orchid Society.

Stone, Maria. 1998. *The Tamiami Trail: A Collection of Stories*. Naples, FL: Butterfly Press.

Svoboda, Albert C., Jr. August 2003. "Conservation in 2003." *Orchids*. Delray Beach, FL: American Orchid Society.

Tisdale, Sally, with photography by Dan Borris. September–October 2000. *Rescuing Orchids*. New York: National Audubon Society.

Yogananda, Paramahansa. 2002. *Inner Reflections: 2002 Engagement Calendar. Selections from the Writings of Paramahansa Yogananda*. Los Angeles: Self-Realization Fellowship.

Acknowledgements

It seems that all of southern Florida helped me to learn, photograph, and write about our state's native orchids. I can only express my greatest gratitude. Some specific individuals and organizations were immeasurably helpful. Paul Martin Brown and John Beckner not only wrote essays but they also tutored me extensively. Dr. Stuart Pimm lent his expertise and prestige. Eric Hansen, a very special friend, inspired me. Thank you to Mike Owen, Karen Relish, and everyone at Fakahatchee Strand Preserve State Park; Lee Hoffman, Tom Coffey, Jim Burch and the members of the Native Orchid Restoration Project, and the staff of Big Cypress National Preserve; Scott Stewart at University of Florida, Gainesville; David Graff at Rookery Bay National Estuarine Research Reserve; Larry Richardson and the staff at the Florida Panther National Wildlife Refuge and Sally Stein at Corkscrew Swamp Sanctuary; Tim Tetzlaff at Caribbean Gardens, the Zoo at Naples; the staff at Marie Selby Botanical Gardens; all the staff and volunteers at the Conservancy of Southwest Florida and Briggs Nature Center; Clyde Butcher, Roger and Lisa Hammer, Mary Ruden, Win Turner, Gary Schmelz, Erik Knudsen, Grant Stephens, and Dave and Judy White. Bob Read and Oscar Thompson, I miss you so much. My daughter, Lea Borkenhagen, Ph.D., encourages her mum, edits kindly, and introduced me to my first orchid love in the rainforest of western Borneo years ago. Ed Bransilver can't tell an orchid from a daisy, but he was patient. David Gerakos, thank you. Finally, thank you to the great inspirations to my photography and my life: Nick Nichols, Joe and Mary Ann McDonald, Art Wolfe, Dewitt Jones, and my friend, the great, great Dr. Jane Goodall, OBE.

International Standard Book Number: 1-56579-501-6

Text and photography copyright: Connie Bransilver, 2004. All rights reserved.

Editor: Elizabeth Train
Designer: Carol Pando
Production Manager: Craig Keyzer

Published by:
Westcliffe Publishers, Inc.
P.O. Box 1261
Englewood, CO 80150
westcliffepublishers.com

Printed in China by C & C Offset Printing, Ltd.

Library of Congress Cataloging-in-Publication Data:
Bransilver, Connie, 1942-
    Wild love affair : essence of Florida's native orchids / photography and text by Connie Bransilver.
        p. cm.
    Includes bibliographical references (p.).
    ISBN 1-56579-501-6
        1. Orchids--Florida. 2. Orchids--Ecology--Florida.
    3. Plant conservation--Florida. I. Title

QK495.O64B694 2004
584'.4'09789--dc22   2003066551

For more information about other fine books and calendars from Westcliffe Publishers, please contact your local bookstore, call us at 1-800-523-3692, write for our free color catalog, or visit us on the Web at **westcliffepublishers.com***.*